Enterprise Artificial Intelligence Transformation

Enterprise Artificial Intelligence Transformation

A Playbook for the Next Generation of Business and Technology Leaders

Rashed Haq

WILEY

Published by John Wiley & Sons, Inc., Hoboken, New Jersey.
Published simultaneously in Canada.

For general information on our other products and services or for technical support, please contact our Customer Care Department within the United States at (800) 762-2974, outside the United States at (317) 572-3993 or fax (317) 572-4002.

Wiley publishes in a variety of print and electronic formats and by print-on-demand. Some material included with standard print versions of this book may not be included in e-books or in print-on-demand. If this book refers to media such as a CD or DVD that is not included in the version you purchased, you may download this material at http://booksupport.wiley.com. For more information about Wiley products, visit www.wiley.com.

Library of Congress Cataloging-in-Publication Data

Names: Haq, Rashed, author.
Title: Enterprise artificial intelligence transformation / Rashed Haq.
Description: First Edition. | Hoboken : Wiley, 2020. | Includes index.
Identifiers: LCCN 2019056747 (print) | LCCN 2019056748 (ebook) | ISBN 9781119665939 (hardback) | ISBN 9781119665861 (adobe pdf) | ISBN 9781119665977 (epub)
Subjects: LCSH: Business enterprises—Technological innovations. | Artificial intelligence—Economic aspects. | Organizational learning. | Organizational effectiveness.
Classification: LCC HD45 .H327 2020 (print) | LCC HD45 (ebook) | DDC 006.3068—dc23
LC record available at https://lccn.loc.gov/2019056747
LC ebook record available at https://lccn.loc.gov/2019056748

Cover Design: Wiley
Cover Image: © Darius Griffin Haq

Printed in the United States of America

V10018487_051320

We are on the brink of the algorithmic enterprise. Today's generation of business and technology leaders can have a metamorphic impact on humanity by catalyzing applied AI for every enterprise.

This book is dedicated to Abbu and Ammi, who encouraged me to pursue my dreams; to Tayyba, who lovingly supported me through life; and to Darius and Athena, who are my endless inspiration.

Contents

Foreword: Artificial Intelligence and the New Generation of Technology Building Blocks

Over the past few years, I have been fortunate to discuss artificial intelligence (AI) with C-suite executives from the largest companies in the world, along with developers and entrepreneurs just getting started in this area. These interactions impressed on me how quickly the conversation is becoming commonplace for business executives, even though AI in business is still in its infancy. As you pick up this book, I hope you realize what an incredible time we live in and how transformative having computers that can mimic cognitive abilities will be in the coming years and decades. Digital transformation is becoming a commodity play as organizations shift into the cloud, and business leaders must plan for and utilize a new set of technology building blocks to help differentiate their companies. Of these, the single biggest impact, in my opinion, will come from AI.

When I entered the software industry over 25 years ago, everything we built revolved around three core elements: computing, storage, and networking, all of which were evolving at an incredible rate:

1. Computing – 286 to 386 to 486 to Pentium, and more
2. Storage – 5¼" floppy to 3½" floppy, to iOmega drives, to thumb drives, and more
3. Networking – corporate network to dial-up modem to DSL, to 2G, 3G, 4G, 5G, and more

The evolution of these building blocks created new computing infrastructure modes (client computing to client-server, to Internet, to cloud and mobile, to today's intelligent cloud and intelligent edge) that allowed technology to better support business and consumer needs. Core computing paradigms and the infrastructure models we all rely on have continued to advance over the past three decades, taking us on a journey in terms of how computer technology is used in business and in our personal lives. Today's conversations have shifted away from traditional technology infrastructures and onto digital transformation and what it means for each business and industry. While the backbone of digital transformation is based on computing, storage, and networking, the next generation is beginning with an entirely new set of building blocks.

These new elements consist of things we read about every day and use in various ways within our organizations: Internet of Things (IoT), blockchain, mixed reality, artificial intelligence, and at some point in the future, quantum computing. The next generation of employees will be natively familiar with these building blocks and be able to harness them to more broadly and dramatically redefine every industry. It is entirely possible that future changes will eclipse the advent of the PC, mobile, and the current round of cloud-driven digital transformation.

Although these building blocks are powerful, AI provides the most potential of any tool to impact businesses and industries.

Unlike the other elements, which apply to clearly defined use patterns, AI can be leveraged in every area of the business. This includes product development, finance, operations, employee management, and supplier/partner/channel alignment. AI can be used to impact both top-line growth and bottom-line efficiencies and leveraged at any point in a business or product lifecycle. Given the breadth of opportunities and the importance of a balanced approach to your organization's AI journey, this book provides a critical reference for business leaders on how to think about your company's – as well as your personal – AI plan.

Each organization will undergo its own AI journey in line with its business strategy and needs. Much like the Internet when it first came along, the excitement and energy for AI is incredibly high and the long-term opportunities immense. Knowing that we are still in the beginning stages of real AI implementations allows us to be more thoughtful and prudent in how we approach this area. Furthermore, the tools and data needed for AI are also on their own journey and continue to evolve at an incredibly high rate.

This move toward production-ready AI is based on three core advancements:

1. Global-scale infrastructure – computing, storage, and networking at scale based on the cloud, which ultimately enables any developer or data scientist, anywhere on the planet, to work with the data and tools necessary to enable AI solutions.
2. Data – the growth of raw data, both machine- and device-driven (PCs, phones, IoT sensors, etc.), and human generated (web search, social media, etc.) provides the fuel for creating AI models.
3. Reusable algorithms – the advancement of reusable models or algorithms for basic cognitive functions (speech, understanding, vision, natural language processing, etc.) democratizes access to AI.

By combining these three elements at scale, any organization or developer can work with AI. Organizations can choose to work at

the most fundamental levels creating their own models and algorithms or take advantage of prebuilt models or tools to build on.
The challenge then becomes where to start and what to focus on.

Today, we are seeing a set of patterns start to emerge within
organizations across a broad set of industries. These include:

- Virtual agents, which interact with employees, customers, and
  partners on behalf of a company. These agents can help answer
  questions, provide support, and become a proactive representative of your company and your brand over time.
- Ambient intelligence, which focuses on tracking people and
  objects in a physical space. In many ways, this is using AI to
  map activity in a physical space to a digital space, and then
  allowing actions on top of the digital graph. Many people will
  think about "pick up and go" retail shopping experiences as
  a prime example, but this pattern is also applicable to safety,
  manufacturing, construction scenarios, business meetings, and
  more.
- AI-assisting professionals, which can be used to help almost
  any professional be more effective. For example, they can help
  the finance department with forecasting, lawyers with writing
  contracts, sellers with opportunity mapping, and more. We also
  see AI assisting doctors in areas such as genomics and public
  health.
- Knowledge management, which takes a custom set of information (e.g., a company's knowledge base) and creates a custom
  graph that allows the data to be navigated much like the web
  today. People will get custom answers to their questions, versus
  a set of links to data. This is a powerful tool for businesses.
- Autonomous systems, which refer to self-driving cars but also
  to robotic process automation and network protection. Threats
  to a network can be hard to identify as they occur and the lag
  before responding can result in considerable damage. Having
  the network automatically respond as a threat is happening can
  minimize risk and free the team to focus on other tasks.

Although these patterns are evolving and do not apply to every business or industry, it is important to note that AI is being used across a variety of business scenarios. So, as a business leader, where do you start? The intent and power of this book are to help business leaders answer this and many other important questions. In this book, Rashed Haq leverages his 20-plus years of experience helping companies navigate large-scale AI and analytics transformations to help you plot your journey and identify where to spend your energy.

A few things to keep in mind as you read this book. The first is that data is the fuel for AI; without data there is no AI, so you must consider which unique data assets your organization has. Is that data accessible and well managed? Do you have a data pipeline for the future? And are you treating data like an asset in your business? The next thing to remember is AI is a tool, and like any other tool it should be applied in areas that help you differentiate as a company or business. Just because you can build a virtual agent, or a knowledge management system, does not mean you should. Will that work help you with your core differentiation? Where do you have unique skills around data, machine learning, or AI? Where should you combine your unique data and skills to enhance your organization's differentiation? At the same time, should you be looking for partners or software providers to infuse their solutions with AI, so you can focus your energy on the things you do uniquely? If you think you have new business opportunities based on AI or data, think about them carefully and whether you can effectively execute against them. Finally, what is your policy around AI and ethics? Have you thought about the questions you will be asked from employees, partners, and customers?

The AI opportunity is both real and a critical part of your current and future planning processes. At the same time, it is still a fast-moving space, and will evolve considerably in the next 5 to 10 years. That means it is critical as a business leader to understand the basics of what AI is, the opportunities it offers, and the right questions to ask your team and partners. This book provides you with the background you need to help you understand the broader AI journey and blaze your own path.

As you begin thinking more deeply about AI and your company's journey, keep this simple thought in mind: "It's too early to do everything . . . it's too late to do nothing" – so leverage this book to help you figure out where to start!

Steve Guggenheimer
Corporate Vice President for AI, Microsoft

Prologue: A Guide to This Book

More business leaders are recognizing the value of leveraging artificial intelligence (AI) within their organizations and moving toward analytical clairvoyance: a state in which they can preemptively assess what situations are likely to arise in their company or business environments and determine how best to respond. The potential for enterprise AI adoption to transform existing businesses to help their customers and suppliers is vast, and there is little question today that AI is an increasingly necessary tool in business strategy. We are on the cusp of creating what has been called the *algorithmic enterprise*: an organization that has industrialized the use of its data and complex mathematical algorithms, such as those used in AI models, to drive competitive advantage by improving business decisions, creating new product lines and services, and automating processes.

However, the whole field of artificial intelligence is both immensely complex and continually evolving. Many businesses are running into challenges incorporating AI within their operating models. The problems come in many forms – change management, technical and algorithmic issues, hiring and talent management, and other organizational challenges. There is emerging legislation designed to protect both data privacy and fair use of algorithms that can prevent an AI solution from being deployed or may create legal problems for companies related to potential discrimination against minorities, women, or other classes of individuals.

Due to these roadblocks, few companies have successfully taken AI into an enterprise-scale capability, and many have not moved beyond the proof-of-concept phase. Scaling AI is a nontrivial proposition. But despite all this, AI is becoming a mainstream business tool. Many startups and the large technology companies are using AI to create new paradigms, business models, and products to benefit everyone. However, the greatest impact from AI will be unleashed when most large or medium-sized companies go through an enterprise AI transformation to improve the lives of their billions of customers. It is an exciting time for today's generation of business and technology leaders because they can have a metamorphic impact on humanity by overcoming the scaling challenges to lead this transformation in their businesses.

I have been lucky to work and talk with leaders in many large organizations as they journey toward incorporating AI across their businesses. The challenges they face are very different from the problems of digitally native companies because they have well-established and successful organizational structures, sales channels, supply chains, and the associated culture. I found that there is a widespread desire for reliable information about applying AI within these organizations but very little literature available that gives a clear, pragmatic guide to building an enterprise AI capability as well as possible business applications. There is no playbook to follow to understand and then address the opportunities of AI. I decided to write this book so that more of today's leaders will understand the appropriate and necessary steps for jump starting a scalable, enterprise-wide AI strategy capable of transforming their business while avoiding the challenges mentioned earlier. This book is the guidebook to help you understand, strategize for, and compete on the AI playing field. This knowledge will help you not only participate but play a leading role in your companies' AI transformation.

The book is a practical guide for business and technology leaders who are passionate about using AI to solve real-world business problems at scale. Executive officers, board members, operations managers, product managers, growth hackers, business strategy managers, product marketing managers, project managers, other company leaders, and anyone else interested in this growing and exciting field will benefit from reading it. No prior knowledge of AI is required. The book will also be useful to the AI practitioner, academic, data analyst, data scientist, and analytics manager who wants to understand how she can deliver AI solutions in the business world and what challenges she needs to address in the process.

I have organized the book into five parts.

In Part I, "A Brief Introduction to Artificial Intelligence," I discuss the different types of AI, such as machine learning, deep learning, and semantic reasoning, and build an understanding of how they work. I also cover the history of AI and what is different now.

In Part II, "Artificial Intelligence in the Enterprise," I cover AI use cases in a variety of industries, from banking to industrial manufacturing. These examples will help you gain an understanding of how AI is already in use today, how it is affecting different business functions, and which of these may apply to your own business to get the most out of your investment. This is not meant to be a comprehensive blueprint of all potential uses within these industries, nor a view of what is possible in the near future.

In Part III, "Building Your Enterprise AI Capability," you will learn what it takes to define and implement an enterprise-wide AI strategy and how to lead successful AI projects to deliver on that strategy. Topics include creating a robust data strategy, understanding the AI lifecycle, knowing what makes a good AI platform architecture, approaches to managing AI model risk and bias, and building an AI center of excellence.

Part IV, "Delving Deeper into AI Architecture and Modeling," will provide a more in-depth description of the architecture, various technical patterns for applications that will be useful as you move further toward implementations, and how AI modeling works using a detailed example.

Finally, Part V, "Looking Ahead," will look at the future of AI and what it might mean for society and work.

Feel free to jump around, reading what you need when you need it. For example, if you are already familiar with AI and understand your use cases, start at Part III. If you are looking for ideas for use cases, take a look at Part II. When you are ready to implement your first set of projects, you can come back to Part IV.

Incorporating AI into your business can be easier than you might think once you have a roadmap, and this book provides you with the right information you need to succeed.

Part I

A Brief Introduction to Artificial Intelligence

Chapter 1
A Revolution in
the Making

The question of whether a computer can think is no more interesting
than the question of whether a submarine can swim.
Edsger W. Dijkstra, professor of
computer science at the University of Texas

Since the 1940s, dramatic technological breakthroughs have not
only made computers an essential and ubiquitous part of our lives
but also made the development of modern AI possible – in fact,
inevitable. All around us, AI is in use in ways that fundamentally
affect the way we function. It has the power to save a great deal of
money, time, and even lives. AI is likely to impact every company's
interactions with its customers profoundly. An effective AI strategy
has become a top priority for most businesses worldwide.

Successful digital personal assistants such as Siri and Alexa have
prompted companies to bring voice-activated helpers to all aspects
of our lives, from streetlights to refrigerators. Companies have
built AI applications of a wide variety and impact, from tools that
help automatically organize photos to AI-driven genomic research
breakthroughs that have led to individualized gene therapies. AI is
becoming so significant that the World Economic Forum[1] is calling
it the fourth industrial revolution.

The Impact of the Four Revolutions

The first three industrial revolutions had impacts well beyond the work environment. They reshaped where and how we live, how we work, and to a large extent, how we think. The World Economic Forum has proposed that the fourth revolution will be no less impactful.

During the first industrial revolution in the eighteenth and nineteenth centuries, the factory replaced the individual at-home manufacturer of everything from clothing to carriages, creating the beginnings of organizational hierarchies. The steam engine was used to scale up these factories, starting the mass urbanization process, causing most people to move from a primarily agrarian and rural way of life to an industrial and urban one.

From the late nineteenth into the early twentieth century, the second industrial revolution was a period in which preexisting industries grew dramatically, with factories transitioning to electric power to enhance mass production. The rise of the steel and oil industries at this time also helped scale urbanization and transportation, with oil replacing coal for the world's navies and global shipping.

The third industrial revolution, also referred to as the digital revolution, was born when technology moved from the analog and mechanical to the digital and electronic. This transition began in the 1950s and is still ongoing. New technology included the mainframe and the personal computer, the Internet, and the smartphone. The digital revolution drove the automation of manufacturing, the creation of mass communications, and a scaling up of the global service industry.

The shift in emphasis from standard information technology (IT) to artificial intelligence is likely to have an even more significant impact on society. This fourth revolution includes a fusion of technologies that blurs the lines between the physical, digital, and biological spheres[2] and is marked by breakthroughs in such fields as robotics, AI, blockchain, nanotechnology, quantum computing, biotechnology, the Internet of Things (IoT), 3D printing, and autonomous vehicles, as well as the combinatorial innovation[3] that merges multiples of these technologies into sophisticated business solutions. Like electricity and IT, AI is considered a general-purpose technology – one that can be applied broadly in many situations that will ultimately affect an entire economy.

In his book *The Fourth Industrial Revolution*, World Economic Forum founder and executive chairman Klaus Schwab says, "Of the many diverse and fascinating challenges we face today, the most intense and important is how to understand and shape the new technology revolution, which entails nothing less than a transformation of humankind. In its scale, scope, and complexity, what I consider to be the fourth industrial revolution is unlike anything humankind has experienced before."[4] This fourth revolution is creating a whole new paradigm that is poised to dramatically change the way we live and work, altering everything from making restaurant reservations to exploring the edges of the universe.

It is also causing a significant shift in the way we do business. Changes over the past 10 years have made this shift inevitable. Companies need to be proactive to stay competitive; those that are not will face more significant hurdles than ever before. And things are happening more quickly than many people realize. The pace of each industrial revolution has dramatically accelerated from its pace in the previous one, and the AI revolution is no exception. Even companies such as Google, which has led the mobile-first world, has substantially shifted gears to stay ahead. As Google CEO Sundar Pichai vowed, "We will move from a mobile-first to an AI-first world."[5]

Richard Foster, of the Yale School of Management, has said that because of new technologies an S&P company is now being replaced almost every two weeks, and the average lifespan of an S&P company has dropped by 75% to 15 years over the past half-century.[6] Even more intriguing is that regardless of how well a company was doing, its prior successes did not afford protection unless it jumped on the technology innovations of the times.

Along similar lines, McKinsey found that the fastest-growing B2B companies "are using advanced analytics to radically improve their sales productivity and drive double-digit sales growth with minimal additions in their sales teams and cost base."[7] In another paper, they estimated that in 2016, $26 billion to $39 billion was invested in AI, and that number is growing.[8] McKinsey posits the reason for this: "Early evidence suggests that AI can deliver real value to serious adopters and can be a powerful force for disruption."[9] Early AI adopters, the study goes on, have higher profit margins, and the gap between them and firms that are not adopting AI enterprise-wide is expected to widen in the future.

All this is good news for businesses that embrace innovation. The changeover to an AI-driven business environment will create big winners among those willing to embrace the AI revolution.

AI Myths and Reality

To most people, AI can seem almost supernatural. But at least for the present, despite its extensive capabilities, AI is more limited than that. Currently, computer scientists group AI into two categories: weak or *narrow AI* and strong AI, also known as *artificial general intelligence (AGI)*. AGI is defined as AI that can replicate the full range of human cognitive abilities and can apply intelligence to any given problem as opposed to just one. Narrow AI can only focus on a specific and narrow task.

When Steven Spielberg created the movie *AI*, he visualized humanoid robots that could do almost everything human beings could. In some instances, they replaced humans altogether. AGI of this type is only hypothetical at this point, and it is unclear if or when we will develop it. Scientists even debate whether AGI is actually achievable and whether the gap between machine and human intelligence can ever be closed. Reasoning, planning, self-awareness: these are characteristics developed by humans when they are as young as two or three; but they remain elusive goals for any modern computer.

No computer in existence today can think like a human, and probably no computer will do so in the near future.[10] Despite the media attention, there is no reason to be concerned that a simulacrum of HAL,[11] from Stanley Kubrick's film *2001,* will turn your corporate life upside-down. On the other hand, artificial intelligence is no longer the stuff of science fiction, and there is already a large variety of successful and pragmatic applications, some of which are covered in Part II. The majority of these are narrow AI, and some, at best, are broad AI. We define *broad AI* as a combination of a number of narrow AI solutions that together give a stronger capability such as autonomous vehicles. None of these are AGI applications.

So how are companies using AI to succeed in this ever-changing world?

The Data and Algorithms Virtuous Cycle

More companies are recognizing that in today's evolving business climate, they will soon be valued not just for their existing businesses but also for the data they own and their algorithmic use of it. Algorithms give data its extrinsic value, and sometimes even its intrinsic value – for example, IoT data is often so voluminous that without complex algorithms, it has no inherent value.

Humans have been analyzing data since the first farmer sold or bartered the first sheaf of grain to her first customer. Individuals, and then companies, continued to generate analytics on their data through the first three industrial revolutions. Data analysis to improve businesses became even more indispensable starting around 1980, when companies began to use their data to improve daily business processes. By the late 1980s, organizations were beginning to measure most business and engineering processes. This inspired Motorola engineer Bill Smith to create a formal technique for measurement in 1986. His technique became known as Six Sigma.

Companies used Six Sigma to identify and optimize variables in manufacturing and business to improve the quality of the output of a process. Relevant data about operations were collected, analyzed to determine cause-and-effect relationships, and then processes were enhanced based on the data analysis. Using Six Sigma meant collecting large amounts of data, but that did not stop an impressive number of companies from doing it. In the 1990s, GE management made Six Sigma central to its business strategy, and within a few years, two-thirds of the Fortune 500 companies had implemented a Six Sigma strategy.

The more data there was, the more people wanted to use it to improve their business processes. The more it helped, the more they were willing to collect data. This feedback loop created a virtuous cycle. This virtuous cycle is how AI works within a data-driven business—collect the data, create models that give insights, and then use these insights to optimize the business. The improved company allows more data collection – for example, from the additional customers or transactions enabled by the more optimized business – allowing more sophisticated and more accurate AI models, which further optimizes the business.

The Ongoing Revolution – Why Now?

Although AI has been around since the 1950s, it is only in the last few years that it has started to make meaningful business impacts. This is due to a particular confluence of Internet-driven data, specialized computational hardware, and maturing algorithms.

The idea of connecting computers over a wide-area network, or Internet, had been born in the 1950s, simultaneous with the electronic computer itself. In the 1960s, one of these wide-area networks was funded and developed by the US Department of Defense and refined in computer science labs located in universities around the country. The first message on one of these networks was sent across what was then known as the ARPANET[12] in 1969, traveling from the University of California, Los Angeles, to Stanford University. Commercial Internet service providers (ISPs) began to emerge in the late 1980s. Protocols for what would become the World Wide Web were developed in the 1980s and 1990s. In 1995, the World Wide Web took off, and online commerce emerged. Companies online started collecting more data than they knew how to utilize.

Businesses had always used internally generated data for data analytics. However, since the beginnings of the Internet, broadband adoption in homes, and the emergence of social media and the smartphone, our digital interactions grew exponentially, creating the era of user-generated data. A proliferation of sensors, such as those that can measure vibrations in machines in an industrial setting, or measure the temperature in consumer products, such as coffeemakers, added to this data trove. It is estimated that there are currently over 100 sensors per person, all enabled to collect data. This data became what we refer to as big data.

Big data encompasses an extraordinary amount of digital information, collected in forms usable by computers: data such as images, videos, shopping records, social network information, browsing profiles, and voice and music files. These vast datasets have resulted from the digitization of additional processes, such as social media interactions and digital marketing. New paradigms had to be developed to handle this Internet-scale data: MapReduce was first used by Google in 2004 and Hadoop by Yahoo in 2006 to store and process these large datasets. Using this data to train AI models has enabled us to get more significant insights at a faster pace, vastly increasing the potential for AI solutions.

Although the volume of data available soared, storage costs plummeted, providing AI with all the raw material it needed to make sophisticated predictions. In the early 2000s, Amazon brought cloud-based computing and storage, making a high-performance computation on large datasets available to IT departments for many businesses. By 2005, the price of storage had dropped 300-fold in 10 years, from approximately $300 to about $1 per gigabyte. In 2010, Microsoft and Google helped further expand storage capacity with their cloud storage and computing-product releases: Microsoft Azure and Google Cloud Platform.

In the 1960s Intel co-founder Gordon Moore predicted that the processing power of computer chips would double approximately every year. Known as Moore's Law, it referred to the exponential growth of the computational power in these computers. In the 1990s, hardware breakthroughs such as the development of the graphics processing unit (GPU) increased computational processing power more than a million-fold,[13] with the ability to execute parallel processing of computations. Initially used for graphics rendering, the GPU would later make it possible to train and run sophisticated AI algorithms that required enormous datasets. More recently Google has introduced the tensor processing unit (TPU) that is an AI-accelerated chip for deep learning computations.

In addition to the hardware, the advances in parallel computing were leveraged to parallelize the training of AI models. Access to these services in the cloud from Amazon, Microsoft, and Google for any company that wanted it made it easier for many companies to venture into this space where they would have been more tentative if each company had to build its own large scalable, parallel processing infrastructures.

Breakthrough techniques in artificial intelligence[14] have been occurring since the 1950s, when early work on AI began to accelerate. Models based on theoretical ideas of how the human brain works, known as *neural networks*, were developed, followed by a variety of other attempts to teach computers to learn for themselves. These *machine learning* (ML) algorithms[15] enabled computers to recognize patterns from data and make predictions based on those patterns, as did the increasingly complex, multilayered neural nets that are used in the type of machine learning known as *deep learning*.[16] Another breakthrough came in the 1980s when the method of *back-propagation* was used to train artificial neural networks, enabling the network to optimize itself without human intervention. Through the 1990s and early 2000s, scientists developed more approaches to building neural networks to solve different types of problems such as image recognition, speech to text, forecasting, and others.

In 2009, American scientist Andrew Ng, then at Google and Stanford University, trained a neural network with 100 million parameters on graphics processing units (GPUs), showing that what might take weeks on CPUs could now be computed in just days.

This implementation showed that powerful algorithms could utilize large available datasets and process them on specialized hardware to train complex machine learning and deep learning algorithms.

The progress in algorithms and technologies has continued, leading to startling advances in the computer's ability to perform complex tasks, ably demonstrated when the program AlphaGo beat the world's top human Go player in 2016.[17] The game of Go has incredibly simple rules, but it is more complicated to play than chess, with more possible board positions than atoms in the universe. This complexity made it impossible to program AlphaGo with decision trees or rules about which move to make when it was in any given board position. To win, AlphaGo had to learn from observing professional games and playing against itself.

The thorny problem of speech recognition was another hard-to-solve need. The infinite variety of human accents and timbres previously sank an array of attempts to make speech comprehensible to computers. However, rather than programming for every conceivable scenario, engineers fed terabytes of data (such as speech samples) to the networks behind advanced voice-recognition-learning algorithms. The machines were then able to use these examples to transcribe the speech. This approach has enabled breakthroughs like Google's, whose Translate app can currently translate over 100 languages. Google has also released headphones that can translate 40 languages in real time.

Beyond speech recognition, companies have now "taught" computers how to both ascertain exactly what a person wants and address that need, all so that Alexa can understand that you want to listen to Bryan Adams, not Ryan Adams, or distinguish between the two Aussie bands Dead Letter Circus and Dead Letter Chorus. Virtual assistants like these can be even more useful, doing everything from taking notes for a physician while she's interacting with a patient to sorting through vast amounts of research data and recommending options for a course of therapy.

Even as technology flashes forward, existing AI techniques are continuing to provide exceptional value, enabling new and exciting ways to conduct tasks such as analyzing images. With digital and smartphone cameras, it is easier than ever to upload pictures to social networks such as Facebook, Pinterest, and Instagram. These images are becoming a larger and larger portion of big data. Their power can be illustrated by research done by Fei-Fei Li, professor of computer science at Stanford University, and the head of machine learning at Google Cloud until recently.

Li, who specializes in computer vision and machine learning, was instrumental in creating the labeled database ImageNet. In 2017, she used labeled data to accurately predict how different neighborhoods would vote based merely on the cars parked on their streets.[18] To do so, she took labeled images of cars from car-sales website Edmunds.com, and using Google Street View, taught a computer to identify which cars were parked on which streets. By comparing this to labeled data from the American Community Survey and presidential election voting data, she and her colleagues were able to find a predictive correlation among cars, demographics, and political persuasion.

Research in AI and its application is growing exponentially. Universities and large technology companies are doing higher volumes of research to advance AI's capabilities and to understand better why AI works as well as it does. The student population studying AI technologies has grown proportionately, and even businesses are setting up AI research groups and multiyear internship programs, such as the AI residency program at Shell.[19] All these investments are continuing to drive the evolution of AI.

This revolution has not yet slowed down. In the past five years, there has been a 300,000× increase in the computational power of AI models.[20] This growth is exponentially faster than Moore's Law, which itself is exponential. However, this revolution is no longer just in the hands of academia and a set of large technology companies. The transition from research to applications is well under way. The combination of current computational power; the enormous storehouse of data that is the Internet; and multiple free, open-source programming frameworks, as well as the availability of easy-to-use software from Google, Microsoft, Amazon, and others is encouraging increasing numbers of businesses to explore AI.

AI: Your Competitive Advantage

Getting value from AI is not just about cutting-edge models or powerful algorithms: it is about deploying these algorithms effectively and getting business adoption for their use. AI is not yet a plug-and-play technology. Although data is a plentiful resource, extracting value from it can be a costly proposition. Businesses must pay for its collection, hosting, cleaning, and maintenance. To take advantage of data, companies need to pay the salaries of data engineers, AI scientists, analysts, and lawyers and security experts to deal with concerns such as the risk of a breach. The upsides, however, can be enormous.

Before AI, phone companies used to look at metrics such as how long it took to install a private line. Hospitals estimated how much money they would bill that would never be collected. Any company that sold something studied its sales cycles – for instance, how long did it take each of their salespeople to close a deal? Using AI, companies can look at data differently. Firms that used to ask "What is our average sales cycle?" are now able to ask "What are the characteristics of the customer or the sales rep who has a shorter sales cycle? What can we predict about the sales cycle for a given customer?" This depth of knowledge brings with it enormous business advantages.

There are undoubtedly potential downsides of trying to use AI applications widely. Building an AI application is complicated, and much of it is utilized without genuinely understanding exactly how it arrives at its decisions. Given this lack of transparency (often called the *black box* problem), it can be difficult to tell if an AI engine is making correct and unbiased judgments. Currently, black box problems primarily involve AI-based operating decisions that appear to handle factors such as race or gender unfairly.

A study by ProPublica[21] of an algorithm designed to predict recidivism (repeated offenses) in prison populations found that black prisoners were far more likely to be flagged as having a higher rate of recurrence than white prisoners. However, when these numbers were compared to actual rates that had occurred over two years in Broward County, Florida, it turned out that the algorithm had been wrong. This discrepancy pointed out a real problem: not only could an algorithm make the wrong predictions, but the lack of algorithm transparency could make it impossible to determine why. Accountability can also be a problem. It is far too easy for people to assume that if the information came from a computer, it must be true. At the same time, if an AI algorithm makes a wrong decision, whose fault is it? Moreover, if you do not think a result is fair or accurate, what is your recourse? These are issues that must be addressed to achieve the benefits of using AI.

JP Morgan's use of AI is an impressive example of how efficient AI can be. The financial giant uses AI software to conduct tasks such as interpreting commercial loan agreements and performing simple, repetitive functions like granting access to software systems and responding to IT requests, and it has plans to automate complex legal filings. According to Bloomberg Markets,[22] this software "does in seconds what took lawyers 360,000 hours."

On the other hand, multinational trading company Cargill is beginning to incorporate AI into its business strategy. In early 2018, the *Financial Times* reported that Cargill was hiring data scientists to figure out how to better utilize the increasing amount of available data. According to the *Times*, "the wider availability of data – from weather patterns to ship movements – has diminished the value of inside knowledge of commodity markets."[23]

Cargill's action illustrates two critical points. Your business strategy may well benefit from using AI, even if you have not yet worked out how to do so. Moreover, given the vast amounts of available data, the current and growing sophistication of AI algorithms, and the track records of successful companies that have adopted AI, there will never be a better time than now to both determine your AI strategy and begin to implement it. This book is designed to help you do both. To begin, we will discuss what AI is and how AI algorithms work.

Notes

1. The World Economic Forum is a Swiss nonprofit foundation best known for an annual meeting that brings together thousands of top business and political leaders, academics, celebrities, and journalists to discuss the most pressing issues facing the world.

2. World Economic Forum (January 14, 2016). The Fourth Industrial Revolution: What It Means, How to Respond. https://www.weforum.org/agenda/2016/01/the-fourth-industrial-revolution-what-it-means-and-how-to-respond/ (accessed September 26, 2019).

3. McKinsey & Company (January 2009). Hal Varian on How the Web Challenges Managers. https://www.mckinsey.com/industries/high-tech/our-insights/hal-varian-on-how-the-web-challenges-managers (accessed September 26, 2019).

4. World Economic Forum (November 27, 2017). The Rise of the Political Entrepreneur and Why We Need More of Them. https://www.weforum.org/agenda/2017/11/the-rise-of-the-political-entrepreneur-and-why-we-need-more-of-them/ (accessed September 26, 2019).

5. *VentureBeat* (May 18, 2017). Google Shifts from Mobile-first to AI-first World. https://venturebeat.com/2017/05/18/ai-weekly-google-shifts-from-mobile-first-to-ai-first-world (accessed September 26, 2019).

6. Innosight (2018). 2018 Corporate Longevity Forecast: Creative Destruction Is Accelerating. https://www.innosight.com/insight/creative-destruction/ (accessed September 26, 2019).

7. McKinsey & Company (January 2018). What the Future Science of B2B Sales Growth Looks Like. https://www.mckinsey.com/business-functions/marketing-and-sales/our-insights/what-the-future-science-of-b2b-sales-growth-looks-like (accessed September 26, 2019).

8. McKinsey & Company (June 2017). Artificial Intelligence: The Next Digital Frontier. www.mckinsey.com/~/media/McKinsey/Industries/Advanced%20Electronics/Our%20Insights/How%20artificial%20intelligence%20can%20deliver%20real%20value%20to%20companies/MGI-Artificial-Intelligence-Discussion-paper.ashx (accessed September 26, 2019).

9. *ComputerWeekly* (June 19, 2017). AI research finds slender user adoption outside tech. www.computerweekly.com/news/450421003/McKinsey-AI-research-finds-slender-user-adoption-outside-tech (accessed September 26, 2019).

10. *VentureBeat* (December 17, 2018). AGI Is Nowhere Close to Being a Reality. https://venturebeat.com/2018/12/17/geoffrey-hinton-and-demis-hassabis-agi-is-nowhere-close-to-being-a-reality/ (accessed September 26, 2019).

11. HAL, incidentally, is a reference to IBM. Each letter in the name of the villainous computer falls right before the letters in the famous tech company.

12. Advanced Research Projects Agency Network.

13. *Soft Computing* 15, no. 8 (August 2011): 1657–1669. Graphics Processing Units and Genetic Programming: An overview. http://citeseerx.ist.psu.edu/viewdoc/download?doi=10.1.1.187.1823&rep=rep1&type=pdf (accessed September 26, 2019).

14. American scientist John McCarthy coined the term in 1955.

15. American scientist Arthur Samuel coined the term in 1958.

16. American scientist Rina Dechter coined the term in the context of machine learning in 1986.

17. The documentary *AlphaGo* (2017) shows how the teams competed in the seven-day tournament in Seoul.

18. *Stanford News* (November 28, 2017). An Artificial Intelligence Algorithm Developed by Stanford Researchers Can Determine a Neighborhood's Political Leanings by Its Cars. https://news.stanford.edu/2017/11/28/neighborhoods-cars-indicate-political-leanings/ (accessed September 26, 2019).

19. Shell: AI Residency Programme – Advancing the Digital Revolution. https://www.shell.com/energy-and-innovation/overcoming-technology-challenges/digital-innovation/artificial-intelligence/advancing-the-digital-revolution.html (accessed September 26, 2019).

20. OpenAI Blog (May 16, 2018). AI and Compute. https://openai.com/blog/ai-and-compute/ (accessed September 26, 2019).

21. ProPublica (May 23, 2016). How We Analyzed the COMPAS Recidivism Algorithm. https://www.propublica.org/article/how-we-analyzed-the-compas-recidivism-algorithm (accessed September 26, 2019).

22. Bloomberg (February 28, 2017). JPMorgan Software Does in Seconds What Took Lawyers 360,000 Hours. https://www.bloomberg.com/news/articles/2017-02-28/jpmorgan-marshals-an-army-of-developers-to-automate-high-finance (accessed September 26, 2019).

23. *Financial Times* (January 28, 2018). Cargill Hunts for Scientists to Use AI and Sharpen Trade Edge. https://www.ft.com/content/72bcbbb2-020d-11e8-9650-9c0ad2d7c5b5 (accessed September 26, 2019).

Chapter 2
What Is AI and How Does It Work?

Early AI was mainly based on logic. You're trying to make computers that reason like people. The second route is from biology: You're trying to make computers that can perceive and act and adapt like animals.
Geoffrey Hinton, professor of computer science at the University of Toronto

The concept of AI is not new. Humans have imagined machines that can compute since ancient times, and the idea has persisted through the Middle Ages and beyond. In 1804, Joseph-Marie Jacquard actually created a loom that was "programmed" to create woven fabrics using up to 2,000 punch cards. The machine could not only replace weavers, but also make patterns that might take humans months to complete, and it could replicate them perfectly.

However, it was not until the late twentieth century that AI began to look like an achievable goal. Even today, artificial intelligence is not a precisely defined term. In an article published on February 14, 2018, Forbes offered six definitions, the first derived from *The English Oxford Living Dictionary*: "The theory and development of computer systems able to perform tasks normally requiring human intelligence, such as visual perception, speech recognition, decision-making, and translation between languages."[1] This is a reasonable place to start because the examples in the definition are the type of AI that is currently being utilized: weak or narrow AI.

The Development of Narrow AI

Computers are driven by algorithms: logical steps written in code designed to produce the desired outcome. The earliest "smart computers" used a kind of algorithm called an expert system. An expert system is based on how an expert in a given field would figure out the answer to a given question, generated as a series of rules that the algorithm follows. Between 1959 and the 1980s, computer scientists focused on developing expert systems (we still use some of these today), building them around domain-specific knowledge created by experts in various fields. As long as the problems to be solved involve formal logical rules that are relatively straightforward to express, expert systems are adequate to the task.

IBM's famous chess-playing computer, Deep Blue, was an expert system, and it was successful because of the nature of the game of chess itself. Good chess players consider every possible move in a given situation and try to play out these options in their heads as far forward as they can. "If I move this pawn, what will my opponent do next? What will she do five moves from now?" In essence, that is what Deep Blue was programmed to do. Expert players "taught" Deep Blue's programmers the moves of chess and the strategies that could win games.

Expert systems were the state of the art when, in 1996 and 1997, Deep Blue faced Russian chess grandmaster Gary Kasparov and won the second of their two matches. But there are many things expert systems cannot do, such as understanding speech or vision or reasoning about the physical world. These tasks are far less structured, and they can be very difficult or even impossible to describe or codify into logical steps. To take them on – to really "compete" with humans – computers must do two things that are far more complicated, things that are easy for people to do but difficult for them to explain: computers must be able to learn from experience and build intuition.

Intuition is much more than just a feeling. It is a process the human brain has mastered to make sense of the multitude of extraordinary and often confusing details presented by the world every day. When we see a bird, whether it is red or blue, large or small, head up or feet up, or even half-decomposed, how do we determine that it is a bird? Does it stand on two legs rather than four? People do that, too. Does it lay eggs? So do turtles. How about the fact that it flies? Airplanes fly.

How do humans make the leap from all the information we have in our brains about birds, such as their different sizes, colors, beak shapes, habits, and ways of flying, to a decision that enables us to know that a particular creature belongs in the category called birds? A two-year-old child can do this. However, for computers, it is a challenging task: it is not easy to create a set of rules that can determine when that computer is looking at a picture of a bird.

People tend to simultaneously underestimate the amount of knowledge required to achieve human-like behavior in computers and underappreciate the results of systems that can come close. There are tasks that humans do efficiently, and others that machines excel at, and these tasks tend to be very different. Scientific or technical tasks often have strict, though perhaps sophisticated, sets of rules that guide their behavior. This type of knowledge is relatively easy to codify; it is why some expert systems were viable even 40 years ago.

Computers are also able to deal with extraordinarily large amounts of data. This ability led scientists to wonder: with enough data, would there be a way for computers to reach conclusions based on this information without the equivalent of human intuition? Could they gather and categorize as much knowledge from "experience" as possible without needing experts to tell them how to do things? Could computers learn to learn for themselves from examples?

The World Wide Web was one of the things that made this idea feasible. Scientists suddenly had a massive online data stream they could utilize. If computers could think like people, learning from all of this data to reach conclusions, something approaching AI might be possible. However, development in this area has been limited. There are currently many approaches within AI that can be applied to various problems, but over the past 10 years, machine learning has been the most prevalent. But in the past five years, deep neural networks, a much more sophisticated version of machine learning, have started to supersede human capability in many areas such as image recognition. These neural nets may ultimately be a component of strong AI.

The First Neural Network

Initially, AI was inspired by logic: if this, then that. However, logic is something people learn later in life. For example, three-year-olds are not thinking explicitly logically; they are learning through observing patterns. We do not yet know precisely all aspects of how the brain works, particularly during those early years. However, there are some relevant theoretical models, and computer scientists decided that one of these might be a much better paradigm for AI than logic. That is how artificial neural networks were born.

The current neuroscience theory is that neurons (brain cells in humans and nodes in computer programs) are arranged in layers, with information first passing through the "bottom" layer, then the next, and so on. That information is continually refined as it passes up the chain. A simple artificial neural net (ANN) was conceived as early as the 1940s. In 1957, Cornell University's Frank Rosenblatt built a prototype he called the Perceptron. Comprising only two layers of neurons – the input layer and the output layer – it nevertheless learned to distinguish between cards marked on the left and cards marked on the right. It was the first learning algorithm.

In 1958, Arthur Samuel, a pioneer in the fields of computer science and AI, introduced the term *machine learning* to encompass the variety of ways computers could learn, including using a Perceptron. Since then, the term has been used to refer to various techniques of AI in which machines learn from experience.

Machine Learning

Machine learning is based on algorithms that can learn from data without relying on explicit domain-specific or rules-based programming – that is, programming specifically designed to solve a particular problem. No one explicitly programs the computer or hand codes any logic to enable it to do a specific task. Algorithms are designed to determine or estimate a function that predicts an output given a set of inputs. Machine learning is a useful approach where there is sufficient and accurately representative data available and when it may be difficult or costly to model the domain knowledge by hand. Rather than explicitly "teaching" a system by modeling human-like knowledge, the system is designed to learn from the data.

The goal of machine learning is to learn that function from a large number of historical observations of input values and their corresponding output values and to accurately predict future output values given future input values. This process may sound relatively simple, but these functions can be very complicated, often too complex for humans to derive. The function being estimated could be from any process (see Figure 2.1). For example, the process might take an input number and multiply it by 2 and give an output number. Alternatively, the process might take a loan application (data) as input and make a loan decision and provide a label of "approved" or "rejected" as the output. Alternatively, the process might be whatever happens in your head to take an input image and give it a label of "cat" or "not cat" as the output. Based on these inputs and

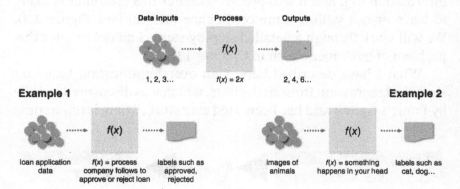

Figure 2.1 Examples of functions $f(x)$ that can be estimated by using machine learning on the input and output datasets.

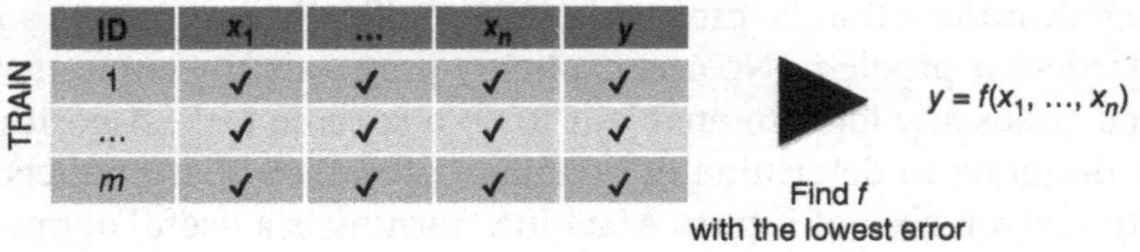

Figure 2.2   Using training data for customers 1 to m to estimate f that will predict y given $x_1, \ldots, x_n$.

outputs, the machine learning algorithm will estimate a function that mimics this process while optimizing for the lowest error.

Let us consider another example. Suppose you had historical data about how a customer behaved. This data might include how many times they called customer support, how much they spent on subscribing to your product or service, and so on. Call these behaviors $x_1, \ldots, x_n$. Suppose you also knew which of these customers closed their account and left (churned) or stayed. Call this y.

Machine learning extracts patterns from the data to look for a function f that most accurately predicts y from $x_1, \ldots, x_n$. This function will be the estimated function with the lowest error, for example, the f with the lowest percentage of false positives and false negatives. This f is your machine learning model (see Figure 2.2). As you get information $x_1, \ldots, x_n$ on new customers, you can pass that information to f, and it will predict whether this customer is likely to leave or not within some confidence interval (see Figure 2.3). We will work through a detailed step-by-step AI model to solve this problem of customer churn in Chapter 13.

What I have described here is an oversimplification based on statistical regression from sixth grade, which was discovered in 1889 by Francis Galton and has been used ever since. There is often some

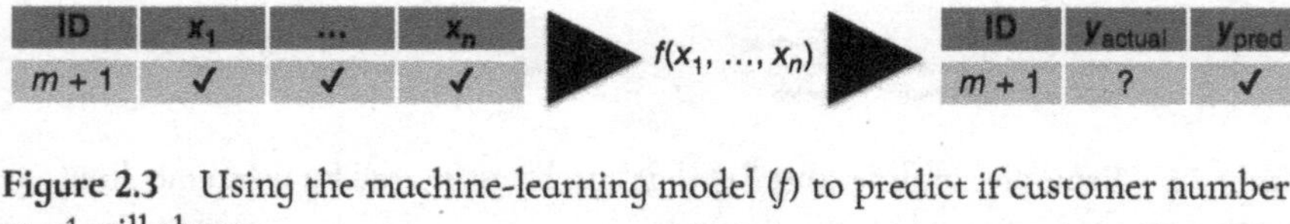

Figure 2.3   Using the machine-learning model (f) to predict if customer number $m + 1$ will churn.

confusion about how machine learning is different from statistics or statistical modeling. Machine learning is a form of statistical learning, but there is a crucial difference between them: machine learning models are for making predictions about future (yet unseen) data, whereas statistical models explain the relationship between historical input data and outcome variables (they are meant to be descriptive and do not make predictions about future datasets). Statistical models look backward, whereas machine learning models look to the future. The similarities arise because both utilize the fundamental concepts of probability. We will explore this concept of "yet unseen data to be able to predict future outcomes" in Chapter 8. Machine learning often also has extreme nonlinearity built into it in the neural network. In our preceding example, if you replace f with a deep neural network with 100,000 parameters to optimize, you get deep learning. Moreover, $x_1, \ldots, x_n$ can be any data, such as the intensity of each pixel in the camera of a self-driving car, and y could be the label "stop sign."

Types of Uses for Machine Learning

Machine learning enables users to organize their data effectively and to derive predictions from it. The vast majority of problems for which machine learning is effective consist of the following three categories.

Classification is the process of predicting the best category for a given new data input, from a predetermined set of categories. Classification is used if the outputs comprise a set of fixed classes – for example categorizing loan applications into those that should be approved and those that should not. The model is trained to classify new inputs into one of these two classes.

Clustering is the process of finding groupings or structures within the data, partitioning the data into clusters that share specific properties that are not from a predetermined set but emerge from the data. This method is often used in customer segmentation to understand their preferences from their profile information and online behaviors.

Regression is the process of predicting a continuous output value corresponding to a given new data input. An example is predicting tomorrow's temperature, which could be a sweltering 98.1 degrees, 98.2 degrees, or even 98.123 degrees. Temperature prediction involves an infinite number of possible outcomes. Unlike in classification, the point is not to predict in what class the new data belongs.

Types of Machine Learning Algorithms

To do classification, clustering, and regression, machine learning uses a variety of techniques or algorithms. The examples that follow are only a few of the methodologies that are currently available. For managers, it is useful to understand the basics about some of the algorithms rather than to become an expert in each to build the intuition of what is going on in the modeling process. This helps with making decisions based on the output of the models because you have a better sense of how the model came to its conclusion. New algorithms are continually being developed, and this is something with which the AI scientists in the organization will need to stay current.

Decision trees are ways to analyze data streams that involve creating a branching system, the nodes of which generally divide data into two buckets: for example, people who liked a movie and people who did not. Subsequent nodes similarly divide the data, growing the tree, branch by branch. There is even something called a data stump, which asks only one question of your data. A variation of decision trees is random forests, and these are made up of many decision trees in which a weighted average is used across the different decision trees.

Logistic regression is used for classification (not regression). Suppose you have a case with two features, such as number of times the customer has complained and their average monthly bill amount. One way to think about logistic regression is that it consists of first plotting data points on a graph with complaints on one axis and bill amount on the other axis, and then finding a straight line that separates the data points into two buckets – with all the data points on one side of the line belonging to one category (for example, "likely to churn") and all the data points on the other side of the line belonging to another (for example, "not likely to churn"). The better the position of the line dividing the two groups, the more accurate your classification. With more than two features, the concept remains similar, and more dimensions are used.

Support vector machines also look for lines that separate the data into two categories in two dimensions, but these lines do not have to be straight; in three or more dimensions, this visualization gets more complicated, but again, the concept remains similar.

Ensemble models are those in which multiple models are trained and used together. A simple approach to this could be to train slightly different models and use the average of the outputs of all of them for a given input. Ensembles often provide higher average accuracy for new data. There are other approaches to ensemble models, including *bagging* and *boosting*. Bagging trains identical algorithms, such as random forest, on various data subsets and applies the ensemble of these to the full dataset. Boosting trains one model after another; each successive model focuses on learning from the shortcomings of the model that preceded it. For example, the second model will focus on predicting for the data inputs where the output was wrong in the first model. In Chapter 13 we will see the kind of accuracy this provides when we switch from a logistic regression model to an extreme gradient boosted model.

Deep learning is an evolution of Frank Rosenblatt's Perceptron. A Perceptron has only one layer of neurons. AI pioneer Marvin Minsky showed in 1969 that there are some useful functions that the Perceptron could never learn. Minsky believed that multilayered, or deep, neural networks might work better, resulting in a multilayer perceptron (MLP). Two researchers at the University of Toronto, Geoffrey Hinton and Yann LeCun, agreed with Minsky's theory. In the 1980s, they theorized that more layers of neural nodes were the key to deeper levels of perception.

The "deep" in *deep learning* refers not to the depth of the computer's real understanding of the data (or the domain knowledge implied by the data) but to the structure of the artificial neural network itself. In a deep neural net, there are more layers of nodes between the input nodes and the output nodes (see Figure 2.4). In some cases, there may be hundreds. The advantage of deep learning is that it has exceptionally high expressive power, meaning that it can be trained to learn very complex functions.

To understand how deep neural nets and deep learning work, let us look at the example of determining whether an image is that of a bird or not. Imagine that the information that arrives in the first layer of the neural network, the input layer, is visual, a collection of pixels of different colors and brightness. A later layer of the neural network might detect the edges in each image, finding parts where one side is darker and the other is brighter. This process would provide a rough outline of the shapes in the picture.

The next layer of neurons might take in the first layer's output and learn to detect more complex factors such as where edges meet at a certain angle, say, at corners. A group of these neurons might respond strongly, for example, to the angle of a bird's beak, although the neural net would have no idea that it is looking at a "beak"; the neurons would merely identify a particular pattern.

The next level might find more complicated configurations, such as a grouping of edges arranged circularly. When these neurons activate, they might be responding to the curved head of the

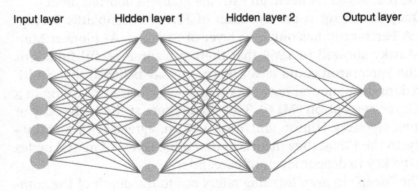

Input layer   Hidden layer 1   Hidden layer 2   Output layer

Figure 2.4   An example of a deep neural network.

bird, although again the neural net would not know the concept of a head. At higher levels, neurons might detect recurring juxtapositions of beaklike angles near headlike circles, which could be a strong signal that the net is "looking at" the head of the bird.

Similarly, the neurons at each subsequent next layer respond to concepts of greater complexity and abstraction, evolving from more straightforward shape recognition toward, for example, the complexity of identifying the outlines of feathers. Finally, the neurons in the top level, or the output layer, correspond to the concepts of "bird" or "not bird." Only one of these will activate in the output layer based on which things triggered in prior layers.

However, if it is to learn, a neural network has to do more than send messages up the layers. It has to determine that it is getting the right results from the top layer based on the labeled data. If this is not the case, the neural network has to send messages back so that all the previous layers can refine their outputs to improve the result. You can think of this as a penalty for not matching the label "bird" for an input image of a bird. This penalty readjusts which nodes in prior layers trigger or do not, based on the input image. If the output layer now matches the label, then it is finished training. If it does not, it causes the network to readjust again, and so on, until the output label matches the input image.

Hence, the deeper the neural nets, that is, the more levels they have, the better the outcome will be up to a certain level of depth. Over time, these systems began to be called deep neural nets, and this concept forms the basis of deep learning. The extra layers in deep neural nets enable an algorithm to discover hierarchical features, features that directly correspond to the inputs, followed by features that correspond to the first-level elements, and so on. This multilevel representation makes deep learning networks better for solving many more types of problems.

Hinton and two of his colleagues wrote a paper offering an algorithmic solution that reduced the errors generated in neural networks. Their approach, known as backpropagation, became the foundation of the second wave of ever-more-accurate neural nets. Backpropagation is a way to more accurately determine how to calibrate the connections between each neuron of one layer to the neurons in higher layers so that the system can come up with increasingly accurate outputs.

Increasingly, neural nets are deep, allowing computers to learn more complicated tasks, such as speech recognition, three-dimensional object recognition, and natural language processing. New deep learning architectures – how the nodes in the neural network are connected to each other across all the layers – are emerging, each suited to different types of problems. Convolutional neural networks (CNNs), for example, are the best at image recognition, and recurrent neural networks (RNNs) are good at natural language processing.

Deep learning models do far better at many problems than do classical (i.e., nondeep) machine learning algorithms. However, they are more difficult to develop, because they require knowledge of neural network architectures and optimization techniques. They also need a lot of computing power and a great deal of training data. Using more straightforward machine learning methods other than deep neural nets may be more efficient or cheaper for some tasks, primarily where they perform well enough, and the incremental model accuracy is not necessary. However, as costs for computing go down, commercial applications for deep learning have increased. Today, for example, deep learning systems are used to look for suspicious banking transactions in real time, triggering text messages about online purchases you might not have conducted yourself.

Supervised, Unsupervised, and Semisupervised Learning

From the 1950s to the 1980s, as algorithms got more powerful, they became capable of learning increasingly more subtle things. The problem was, there was not enough data available for them to make accurate predictions. With the advent of the World Wide Web in the 1990s and digitization of business processes, massive amounts of data began to become available. Researchers began to think about how they might use this new information to enable computers to learn from it. One step they took was to label the data.

Labeled data is data that is already categorized, often by a human being. "Birds" is a category. "Not-birds" is also a category. To train a neural network to recognize birds, you might label a thousand images that contain them as "birds," as well as a thousand pictures that do not include birds as "not-birds." Together, these two sets of labeled data can be used to train a neural network to recognize images of birds.

Using labeled data in this way is known as *supervised learning*. In supervised learning, the training data provided to the system includes the desired outputs, or conclusions, for each input. The data may also include positive and negative examples of conclusions. Supervised learning is currently used for everything from identifying cancerous cells to recognizing spam. A spam filter looks at many factors when sorting through email, such as where the message comes from and what software was used. Spammers tend to use software, or spam engines, that send out a large volume of untraceable messages quickly. Spam filters also search through strings of characters within emails that categorize already-known spam – such as messages about Viagra or requests from a Nigerian prince.

However, as spammers get smarter, simple rule-based spam filters begin to fail. To get ahead of this problem, computer scientists began using supervised machine learning to distinguish spam from not-spam. Scientists already had a host of data about what spam looked like, partly thanks to manually labeled data. So, they developed algorithms to enable a program to learn what was spam and what was not, using existing examples they had of emails that had been categorized as spam or not-spam. Even today, when you mark emails as spam, this labels the data, which is then used to retrain the algorithms.

Identifying cancer is somewhat different from identifying spam, but supervised learning has been used to accomplish this as well. Of course, there are no cancer cell filters on which algorithm development can be based. More importantly, to detect cancer cells, the computer needs to analyze images; and every picture of every cell consists of an enormous number of data points. The machine has to figure out what in the combination of all these data points makes a cell abnormal. Fortunately, human beings can be trained to identify cancer cells, and we already have a large amount of data about these abnormal cells, so it is possible to build an algorithm to enable the computer to learn to tell the difference.

The process begins with feeding large numbers of images of cells into the system. As in the spam model, two datasets are used: one set whose images have been previously determined, most likely by a doctor, to be cancerous, and another set of healthy cells. By learning from this data, the computer can look at an image it has not seen before and classify it as cancerous or noncancerous.

Both the spam and cancer cell algorithms fall into the supervised learning category. In both, previously gathered labeled data exists (or is created) that can be organized and used to enable the computer to recognize patterns in the data. Generally, problems such as cancer detection are too complicated for computers to analyze using unsupervised learning. Most machine learning applications today rely on supervised learning. However, unsupervised learning can be useful in many cases.

In 1997, two computer scientists developed a concrete example of unsupervised learning using clustering.[2] The idea was to help credit-granting businesses predict how to make reliable credit worthiness decisions about a new customer without any prior classification data about the customer and information about previous credit history. Instead, they utilized data about existing customers. The AI scientists created an algorithm that clustered existing customers into groups based on factors such as their use of credit cards and whether they paid off their cards on time. With this information, they created a model that helps these companies know where a customer would fall within various credit worthiness groups at the time of enrollment, without access to the specific credit rating of any given customer.

Grouping unlabeled data, in which the computer is asked to determine categories without a human having looked at the data beforehand, is called *unsupervised learning*. In unsupervised learning, the desired outputs are not provided. The goal for unsupervised learning is for the computer to uncover inherent structures and patterns in the data, or relationships among the inputs. Typically, unsupervised learning is used to identify clusters in the existing data, enabling it to categorize new input data into one of these clusters. Unsupervised learning is also used for anomaly detection, in which inputs that are outliers – i.e. that do not fall into groups where most of the data lie – can be identified. Anomaly detection might be used in manufacturing to determine if a part has the desired shape or not, for example, to check if a tooth in a gear has been chipped.

An additional category known as *semisupervised learning* utilizes techniques employed in both supervised and unsupervised learning. Semi supervised learning is often used when the labels for a dataset are incomplete or error prone. In semisupervised learning, the algorithm employs clusters, as is done in unsupervised learning, but it also uses small labeled datasets. That way, when these labeled sets show up in a particular cluster, the algorithm has additional information about the nature of that cluster. The algorithm will not operate with perfect certainty, but it can "recognize" the likelihood that other samples in the same group should be labeled similarly. Of course, semisupervised learning requires manual reviewing and analyzing to validate or invalidate the resulting models. Today it is being used in anti-money laundering and fraud detection applications.

Sidebar: ImageNet

Both supervised and unsupervised learning are in wide use today. Each can be and will continue to be applicable under certain circumstances. However, it was supervised learning that would generate one of the big developments in AI. In 2009, an extensive database of images was made available online by researchers from the computer science department at Princeton.[3] A year later, a contest was launched:

the ImageNet Large Scale Visual Recognition Challenge (ILSVRC). The goal was to see how many images an algorithm could correctly categorize: that is, how many photos it would classify correctly. Pictures of cats had to go in the cat category, images of dogs into the dog category, and so forth.

Four computer scientists at Stanford, two of whom had been Hinton's students, entered the contest in 2012, applying a deep neural network architecture they had developed to the ImageNet database. Their software identified objects with twice the accuracy of their nearest competitor. The improvement was an astounding development, providing conclusive proof that a deep neural network could work significantly better than any previous AI method. Three years later, in the same contest, the Stanford team's new AI algorithm surpassed human performance in the identification task for the first time. In the future, ImageNet may be considered today's version of the Rosetta Stone for computer vision.

At Google, Ng was involved in another groundbreaking experiment. His team exposed large neural networks to 10 million unlabeled thumbnails of videos from YouTube and then let their algorithm learn how to identify cats in an unsupervised way. When they tested the algorithm with new data, it correctly identified cats 74.8% of the time. Unsupervised learning had never been used on such a scale before and with this degree of success.

Making Data More Useful

A computer's ability to learn hinges not only on how much data is available. It also depends heavily on how the data is represented. That means for a computer to identify a bird accurately as a bird, whether it is a crow, an owl, a pelican, an egret, or a chicken, it must "know" whether the image it is seeing is right-side-up or upside-down or in fog or snow or sun or shade. It must know the picture of a bird is a bird even if only part of that bird is showing, maybe as little as a beak. It must exclude photos of anything that might look like a bird, such as a feathered headdress or a feather duster. This analysis is not easy, which is why (as of this writing) online tests to determine if you are human include various incomplete images of things such as cars or road signs.

To compensate for some of the challenges that computers face in data representation, computer scientists developed *feature engineering*. A feature is an attribute or property of the data on which the computer hinges its analysis. Feature engineering is designed to choose the best features of the data, creating the best representation of data so machines can learn more efficiently. Initially, this task was difficult and expensive, mostly because it had to be done "by hand" by AI scientists. Currently, there are a variety of automatic feature engineering methods that facilitate the tweaking of data, an essential development because, regardless of the time or cost, feature engineering is often necessary to work with data effectively.

As an illustration of the necessity for this, consider the difference between Roman and Arabic numerals – e.g. 10 versus X. If a problem involves numerical representations, say, adding a column of 10 large numbers, which representation of numbers would be an ideal choice: Roman or Arabic? Which would make it easier, in fact, possible, for the calculation to be done by a human? And for a computer to add those numbers, it is easier to convert to binary representation – so a 10 would be 1010. Feature engineering helps create the right representation of data for machine learning algorithms in a similar vein, so the algorithms can perform the computational learning they are tasked with.

Beyond feature engineering, *dimensionality reduction* is a way to choose the features that matter most from all available features, ignoring those features that might not be the ideal predictors of an outcome. For example, if someone wanted to predict how likely it was for a potential customer to purchase a new product, knowing that customer's name would not lead you to a useful outcome, but knowing how much they have spent on the product in the past likely would.

Feature engineering is less necessary for deep learning models. Deep learning generates the data representation, relieving humans of having to do feature engineering. These models use sufficient training data to figure out what in the data leads to the correct answer. It is worth remembering, as noted earlier, that features used in deep learning do not necessarily have any clear concepts that correspond to human-friendly ones. For instance, that deep learning neural net that recognized images of birds may not have any features such as heads or beaks. If you were able to ask one of these systems, "What made you decide that this was a picture of a bird?," it might respond with an answer that is not human interpretable in terms of features of birds.

That would not be an illuminating answer, so learning to use systems like these within a business can be a little tricky. It is trickier still to justify why individual decisions were made by a deep learning model, especially in regulated or customer-facing situations such as when a customer wants to know why she was turned down for a mortgage loan. These systems are often referred to as black boxes, and they can cause a variety of problems. (We will cover how to mitigate for the black box problem in Chapter 10.) Building a system that would be able to say "I came up with 'cat' because it has ears, eyes, a tail, fur, and paws" would have real advantages. One type of AI system that can do this is currently using a different kind of AI than machine learning: semantic modeling and causal reasoning, or semantic reasoning for short.

Semantic Reasoning

Machine learning models arrive at their conclusions by recognizing patterns or correlations between inputs and outputs. They answer questions such as "Of all the cells, which are cancerous?" or "Of all the email, which are spam?" by learning patterns from labeled data and predicting outcomes for unlabeled data. In certain situations, however, these systems are less than ideal. Machine learning systems need a considerable amount of data to train their algorithms – data that may not be available. For example, a machine learning model employed to match outcomes from various medical protocols to genetic anomalies in individual patients would need a great many patients with those anomalies to achieve meaningful results, and there might not be that many patients available. Semantic reasoning systems can, in part, overcome this difficulty.

Semantic reasoning models do not require the same quantity of labeled data as machine learning approaches, although they do need an ontologist who has a deep understanding of the information and rules that need to be captured. They are rules-based systems that can have specific information about entities and their relationships, and they utilize this information to make inferences. In semantic reasoning, the world is described through a set of concepts, their descriptions, and the relationships among them. Semantic reasoning systems make inferences from causes to effects, enabling them to infer logical consequences from a set of given facts and concepts. The knowledge model and inference engine allow these systems to model knowledge with higher fidelity and support more sophisticated reasoning when answering questions.

To illustrate how this works, we can take the sentence "John watched *Mission Impossible* yesterday" and convert it into a semantic model (see Figure 2.5) of entities or concepts, attributes, and relationship. Entities represent something in the real world, e.g. a person. Similarly, concepts represent an abstract idea that is not an entity, e.g. a time, an activity. Attributes represent something about an object or concept, such as a name or an age. Relationships connect different entities and concepts.

Semantic reasoning can generalize beyond the collected data. For example, it may have concepts such as cat and mammal, and facts such as "Sylvester is a black and white cat." Then it can infer things, such as the fact that Sylvester likely has a tail and two ears. To teach a machine learning model to recognize, say, under what conditions a human gene might turn off so someone could create a drug that would prevent this from happening would be pragmatically unfeasible using machine learning alone, because machine learning is based on recognizing patterns or correlations, not on causality, and the correlations are sometimes not easily interpretable by human beings.

A biological pathway is a series of causal interactions among molecules that lead to a change in a cell, such as turning a gene on and off. Machine learning might enable you to look at a large amount of data to find which molecular interactions are most common when a particular gene is turned off, but it would not tell you which interactions turned off the gene. Machine learning can discover correlations but not causality. Because of the hundreds of thousands of correlated molecular interactions that occur and without knowing the exact causal pathway, it would be prohibitively costly to create a drug to prevent the gene from getting turned off using just machine learning. Instead, it would require an AI system that can handle semantic reasoning.

It is clear why teaching a computer to use semantic reasoning could be very useful. It would obviate the need for a ton of data or massive amounts of computer time, enabling a host of applications, from gene therapy to understanding what made a marketing

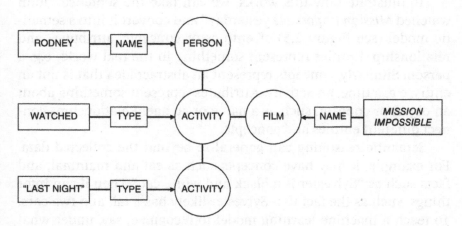

Figure 2.5   Example of a type of knowledge graph.

campaign go viral. Existing machine learning models cannot do that. It could tell you, however, whether there are patterns among the viral campaigns versus the ones that did not go viral. Also, there are other areas in which semantic reasoning could improve outcomes, for example, in the field of law enforcement, using location data from crime scenes and surrounding areas, the context in which a crime is committed, sophisticated knowledge of vocabulary, and context (the difference, for example, between arrest in police jargon and arrest in the medical field). The subtleties of this and other areas will vastly benefit from semantic knowledge modeling.[4]

Thus far, there have been only rudimentary successes in the field. Semantic reasoning made some progress in the 1980s and 1990s but has had minimal success, whereas the most prominent recent AI advances continue to come from deep learning, which is why most of this book will focus on machine learning and deep learning. However, although semantic reasoning is generally neither scalable nor flexible, research has indicated that it might be used along with machine learning and deep learning to enable AI reasoning. As of now, a combination of deep learning and the current state of semantic reasoning may prove to be the most useful approach.[5]

Today's systems do exceptionally well in predicting from existing data over specific domains. They have recently been used quite successfully, for example, in cybersecurity, in which algorithms studied the code in computers and were able to figure out where vulnerabilities might lie and how they might be fixed. However, machine learning systems are not very good at abstraction: applying the knowledge that they have learned and using it on another level. To do this, researchers may have to take a hybrid approach. This approach requires two key components: a broad, deep, and high-fidelity computer-usable model of the relevant domain and background knowledge (the data), as well as a reasoning (inference) engine capable of efficiently combining this knowledge to answer a question or reach a conclusion.

Currently, although some aspects of a given knowledge model can be inferred, much of the knowledge content must be manually curated, requiring ongoing improvements, which can be extremely expensive. These models also may rely on individual expert opinions, which could be incomplete or even incorrect. Available semantic software packages vary in their representational capability and fidelity as well as in the extent of their knowledge bases or the amounts of data they need as a starting point for subsequent knowledge modeling. These software packages provide a basis for doing some reasoning but are limited in representational power, as well as in the domain knowledge they must build on.

Another path toward semantic reasoning may come from deep learning. Although it is too early to know if this will work out, it remains an active area of academic research. In just the past few years, major strides have been made in the field of natural language processing. There are now language models that not only do vastly better data mining than before, but also seem to simulate very rudimentary types of reasoning. The *New York Times* recently reported

Sidebar: Cyc and ConceptNet

In the early 1990s, scientists began developing new ways that data and relationships among data elements could be represented. These were called representation frameworks. At one end of the spectrum were languages such as Unified Modeling Language (UML) and Web Ontology Language (RDF/OWL), which was based on a World Wide Web standard known as the Resource Description Framework, which is one of these representational frameworks. Each of these general-purpose, developmental modeling frameworks was meant to provide a standard way of representing the design of a system. They might include things like the individual components of a system (you can think of this as the static view, or the nouns, of the system) and how they might interact with others (a dynamic view, or the verbs, in the system).

At the other end of the spectrum is the data dependent Cyc platform (created by Cycorp®). Cyc is the world's longest-living AI project, an attempt to create the largest-ever gathering of data that explains how the world works but focused on things that are rarely written down or said, meaning they would not be a "standard" data stream. This amount of data presumably would allow the program to be more flexible within new and unanticipated situations.

Cyc provides an existing knowledge base containing hundreds of thousands of concepts interrelated via tens of thousands of relation types. It also provides a powerful and efficient reasoning engine that supports both forward and backward inference as well as deductive, inductive, and abductive reasoning. The Cyc knowledge base and reasoning engine enable it to provide a higher-order logic modeling language that supports features such as higher-parity relationships (i.e. not limited to subject-verb-object statements), and the ability to make assertions (statements) about other assertions: a powerful mechanism for putting knowledge in context.

Another approach to creating a broad base of human knowledge for computers to work on is the Open Mind project from MIT, which crowd sourced a collection of millions of facts about the world in English rather than in a formal computer-usable representation. The idea was to consolidate these bits of common knowledge, which comprised, MIT posited, what we call "common sense," and convert them into a computer-usable form. This knowledge was (partially) captured in ConceptNet, an extensive graph of interrelated concepts that can support some tasks, such as semantic search. Semantic search refers to improving search accuracy by understanding the searcher's intent using the contextual meaning of terms in the data space.

that for the first time, an AI system was able to pass both an 8th-grade and a 12th-grade science test.[6] Aristo, as the system is known, shows the progress in this area.

Applications of AI

At a high level, how businesses are applying AI can be grouped into three areas. Some companies are *eliminating repetitive tasks* for employees – liberating workers from repetitive, mundane knowledge tasks that require little cognitive effort. This automation improves both cost savings and data accuracy and is often done using robotic process automation (RPA), which uses rules to emulate what a user would do on keyboards across various business apps.

Other companies are *generating insights* – extracting actionable, nontrivial, and previously unknown knowledge and insights from structured and unstructured data. This is often done using machine learning and deep learning. These insights are then acted on to get value from them. Still others are *augmenting human intelligence* by providing contextual knowledge and support, to help customers and employees perform tasks in increasingly more straightforward and more effective ways. This is often done using virtual assistants or processes built into the appropriate context within existing applications and workflows.

You can think of these three applications of AI as falling into the categories of machines that act, machines that learn, and machines that reason (see Figure 2.6). Today, there are very few systems that can come close to reasoning except in a trivial sense. Over the past five years, many innovations in AI, machine learning, and deep learning have made their way into businesses. Moreover, the majority of the enterprise adoption has been in machines that act (e.g. RPA) and machines that learn.

In Part II of the book, we will look at some use cases in specific industries to give a sense of the variety of AI applications that are being used today. Industry drivers behind the AI use cases we will see include reducing costs by automating tasks, reducing risk

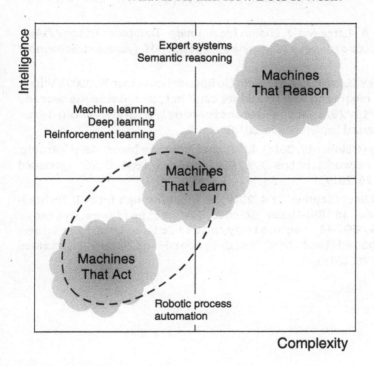

Figure 2.6 Types of AI systems.

through more accurate predictions, improving customer service, achieving compliance with a variety of national and international regulations, and increasing revenue.

Notes

1. *Forbes* (February 14, 2018). The Key Definitions of Artificial Intelligence (AI) That Explain Its Importance. https://www.forbes.com/sites/bernardmarr/2018/02/14/the-key-definitions-of-artificial-intelligence-ai-that-explain-its-importance/#37002fab4f5d (accessed September 26, 2019).
2. AAAI (1997) Clustering and Prediction for Credit Line Optimization. https://www.aaai.org/Papers/Workshops/1997/WS-97-07/WS97-07-006.pdf (accessed September 26, 2019).

3. ImageNet: A Large-Scale Hierarchical Image Database. http://www.image-net.org/papers/imagenet_cvpr09.pdf (accessed September 26, 2019).

4. Fern Halper's Data Makes the World Go Round (November 29, 2007). What's a semantic model and why should we care? https://datamakesworld.com/2007/11/29/whats-a-semantic-model-and-why-should-we-care/ (accessed September 26, 2019).

5. Arxiv.org (October 17, 2018). Relational inductive biases, deep learning, and graph networks. https://arxiv.org/abs/1806.01261 (accessed September 26, 2019).

6. *New York Times* (September 4, 2019). A Breakthrough for A.I. Technology: Passing an 8th-Grade Science Test. https://www.nytimes.com/2019/09/04/technology/artificial-intelligence-aristo-passed-test.html?smid=nytcore-ios-share (accessed September 26, 2019).

Part II

Artificial Intelligence in the Enterprise

Chapter 3
AI in E-Commerce and Retail

Customers will fulfill their everyday needs – items like laundry detergent, paper, light bulbs, grocery staples and shampoo – in the easiest way possible through a combination of stores, e-commerce, pick-up, delivery and supported by artificial intelligence.

Doug McMillon, president and CEO, Walmart

E-commerce and brick-and-mortar retail businesses are currently leveraging the insights gained from sophisticated AI algorithms to improve and grow their enterprises. This is leading to productive collaboration among sales, customer service, and advertising and marketing divisions, which are increasingly sharing data, AI platforms, and analytics teams. Data from advertising, marketing, customer transactions, and customer service is also being integrated into supply-chain functions to improve demand forecasting and fulfillment, as well as returns optimization. Over time, there will likely be more significant integration of these functions. The following are some of the areas in which AI models are successfully being used in e-commerce and retail businesses.

Digital Advertising

Probably the largest success of AI use at scale is in digital advertising and marketing. In 2019, for the first time, businesses will spend more money on digital advertising and marketing than on traditional media, such as TV, radio, and newsprint.[1] Many AI innovations have come from the largest advertisers – Google, Facebook, Alibaba, and Amazon. AI algorithms have helped companies that utilize digital advertising in its various forms: sponsored search advertising, contextual advertising, display advertising, and real-time bidding auctions. AI models accurately and quickly predict ad click-through rates based on customer segments, which creates the basis for the value of the ad. Although much of the focus has been on digital advertising, traditional advertising has also benefited from AI – for example, some companies are using AI-based customer segmentation models from digital ads to help inform traditional advertising such as TV ads.

There are two primary ways to advertise online: during a search, when the user chooses to see ads, and on websites and mobile apps. When someone uses a search engine, the search engine matches a set of possible ads to that search based on specific, advertiser-chosen keywords. An auction mechanism then chooses which ads to show to that user and determines the rate the advertiser will pay to the search engine for displaying the ads. Being able to predict if an ad that is displayed will be clicked on makes AI an essential component of this auction because AI models have better predictive capability than any methods previously used. Choosing which ads to feature on a website works in a similar way, but rather than associating an ad with a search query, ads are targeted at users browsing the site, based on their demographics and other information.

Real-time digital advertising involves multiple actors: the publisher, such as an online car magazine that wants to sell ad space on their website; the advertiser, such as a bank that wants to show an ad for a car loan; and a series of service providers that help to do the matching. The publisher commissions a supply side platform (SSP) to offer advertising space to an ad exchange. The ad exchange acts as the marketplace in which supply-demand (or space-to-ad) matching happens. Advertisers use demand side platforms (DSPs) to compete for and bid on available ad spaces, employing AI to achieve high click-through rates and conversion rates. Data management platforms (DMPs) offer individual viewer profiles and sell interest data to support decisions in the auction process.

In real-time digital advertising, individual advertising spaces or slots are sold within a few milliseconds after the viewer clicks on a website: for example, when a viewer clicks on a car magazine site. As the webpage is loading from the server, a request is made to the SSP for the slot where the ad needs to be shown. The SSP uses the specifications of the ad space, such as size and position of the slot, minimum price, what kinds of ads are allowed to be shown, and any relevant user information, to trigger an auction on the ad exchange. The ad exchange determines which DSPs this slot may be appropriate for and forwards them an auction bid request. The DSPs then decide which of their advertisers' campaigns are relevant. They usually consider all the available information about the current viewer. If there is a cookie, device, or other identifier of the user, the DSP looks up additional information about the user from a DMP, such as the user's interests or demographics. This information is then used to make decisions in the bidding process.

When the ad exchange receives all the bids, it awards a winner. The winning DSP sends the location of the ad server for the media, such as the ad image or video, to the exchange. This information is then sent back to the webpage, which is still loading, since all this activity has taken only milliseconds. The ad server serves the media and tracks all subsequent user activity related to the ad materials. The DSP also uses this user tracking information for future ads on which it will bid.

The use of AI models enables advertisers to make decisions about whether to bid on an ad space based on predicted user responses. This is why most online advertisers use click-prediction systems. Predictions are often based on customer segmentation of the user, which is done using machine-learning-based clustering. AI is particularly useful in the digital ad space because customer behavior is very diverse and robust, and adaptive AI algorithms are uniquely able to predict behaviors based on models learned from historical data.

Marketing and Customer Acquisition

Retail e-commerce is growing at a rapid clip; it exceeded $500 billion in 2018.[2] A lot of this growth is due to AI and big data. Some of the uses of AI in marketing to acquire new customers include *AI-driven marketing campaigns* to target prospects, and uplift and attribution modeling to optimize marketing spending.

In the past, determining which prospective customers browsing an e-commerce site would convert and generate revenue was somewhat hit or miss. If the prospect left the site without buying, it was hard to bring her back into the purchase funnel. Commonly, generic retargeting has been used, in which ads for a retailer's products are shown to all individuals who had interacted with the retailer's site. Today, however, AI-enabled "prospect sensing" technology and AI-driven marketing is making targeting and retargeting more precise – not just in retail, but for e-commerce across industries.

The key to success in prospect sensing is the ability to collect prospect and customer data and act on that data appropriately. We refer to someone as a *prospect* if we do not know who they are but know that they may be a potential customer; a *customer* is someone who is either logged in or who has purchased from the retailer in the past. For prospects, most online interactions, such as browsing or digital advertising, are tracked through either a device identifier (ID) on a mobile app or another tracking resource such as a logged in identifier, cookie, script, pixel, or image on a browser. In a *New York Times* opinion piece, Farhad Manjoo discussed his exploration of how much digital tracking happens through websites and found "everything you do online is logged in obscene detail" and he "was bowled over by the scale and detail of the tracking."[3] The tracking data was beautifully illustrated by Nadieh Bremer.

If the user is an existing customer, most e-commerce companies will keep the tracking ID in the customer database, together with email and other information. Even if the customer is browsing without logging in, their current tracking IDs can be matched to the ones in the customer relationship management (CRM) system to know who they are, and the site or app can be personalized based on this information. If the user is a prospect who is browsing the business's e-commerce app or website anonymously, only the prior browsing information associated with her tracking ID is available to use. This tracking ID is crucial in collecting enough data to learn and predict the behaviors of the prospect or the customer. The tracking ID is also used in serving digital ads to that prospect's device or browser, without needing any personally identifiable information (PII).

The types of data collected when a prospect uses an app or visits a site include data about where the user came from (for instance, from an ad or by using a search engine) and what she did while she was on the site. If an ad was served on a smartphone, her location information can be collected, if she has granted access to the app or browser to enable this tracking. Some of these prospects eventually become customers, leaving digital breadcrumbs of their path to conversion.

The conversion event itself – that is, whether a purchase was made – can be thought of as a label for supervised machine learning. Companies use the relevant data to produce comprehensive timelines of interactions with each customer, known as the *customer journey*. This journey map enables vastly improved predictive ability and new and productive ways to deal with prospects and customers – for example, determining not just whom to target but when and through which channel to communicate. Critical to helping reach customers with journey thinking are the AI models behind dynamic content and next best experience.

Many companies leverage this data in their AI-driven marketing campaigns to more accurately predict user behavior, using AI models to create personalized messages and offers, which improve revenue and marketing return on investment (ROI). As a result, marketing is moving away from the mass-market, one-size-fits-all model toward the one-to-one or one-to-you ideal now made possible by AI capability and data availability. For example, e-commerce companies use AI to learn from the data mentioned earlier and, given a new prospect's behavior, are able to predict if that prospect is likely to buy or not. If the potential customer is not likely to convert, then avoiding retargeting saves marketing expenditures. If the potential customer was thought to be likely to buy but does not complete a transaction, she can be retargeted more precisely with ads. In this way AI, can nudge relevant prospects back into the buying journey.

Targeting prospects who have yet to become customers is primarily enabled through third-party and owned ad-impression data at the individual level, such as the historical ad impressions described earlier. Consider a woman who is in the market for a large-screen TV. First, she visits Google and types in the search query "best large-screen TV." Google lists sites that provide information about TVs, such as TechRadar or CNET. Retailers who sell TVs have ads on these sites. By placing the ads, the prospect is tracked with the cookie ID or, if they are on their mobile app, a device ID. Even if the prospect does not click the ad to enter the advertised site, the retailer who won the bid to place the ad will already know something about her because the ad was loaded or viewed. The retailer can now place more personalized ads back on CNET or TechRadar for her to see the next time she is online. If she visits the retailer's e-commerce site, the retailer can personalize the site to highlight TVs and specifically large-screen TVs.

In the absence of individual data, or sometimes in conjunction with it, many companies incorporate other third-party data such as demographics or weather at a customer segment level, including data such as average income level based on zip code. For example, consider a car dealership that is looking to sell cars to the most likely prospects. The dealership has some first-party data about their previous customers' behavior. But there is no information available about potential customers' income or spending habits. To be able to send out targeted flyers offering the most appropriate deals to likely prospects in different parts of town, the dealership acquires publicly available census data. The dealership may even decide to purchase data about the kinds of cars parked in the neighborhoods for which they acquired this census data. This enables the dealership to use AI models to determine which nearby neighborhoods are particularly price sensitive and what makes and models of cars the residents are likely to drive. The dealership then targets potential customers in that area with special deals on particular makes and models of cars.

Companies such as Publicis Epsilon, Acxiom, Liveramp, and Experian provide services that take user data from a business, connect it to other user data from other sites, and give this full dataset back to the company, stripping out any PII. This consolidated dataset enables the retailer to know more about what a prospect browsed and viewed. Of course, "she" is not identified – sometimes referred to as being pseudonymous. However, if her tracking ID shows up in the future, the retailer will surface specific ads tailored to her profile.

In marketing, AI is also used for *uplift modeling* to optimize marketing spend. Uplift modeling determines which customers should be targeted with an offer and which customers should not by predicting the estimated change in conversion rate due to a given campaign. Suppose a retailer is planning a marketing campaign to increase conversion by offering discounts for a product. Each customer gets an uplift score representing the increase in the probability of her making a purchase if she receives the discount. If a customer already has a very high likelihood of buying, the discount will decrease revenue because she will have likely bought in any case. If she has a high uplift score but a low probability of buying otherwise, a discount will increase revenue. If she has a low uplift score, then targeting her with offers and ads will likely be a waste of marketing dollars. If the customer has a negative uplift score, that is, if she is less likely to convert because of the campaign, then targeting this particular customer will lose revenue, both because it will turn her away and because marketing dollars will be spent unnecessarily. In some cases, uplift is used for modeling pricing optimization. For example, many companies give discounts for their products. The budgets for these incentives are often 10 times larger than marketing budgets, and optimizing these have significant impact.

Another aspect of managing marketing budgets is to decide how much to allocate to different campaigns and channels. To do this, marketers must know how much revenue past campaigns and channels generated and attribute revenue generated to their digital marketing spending. It can be difficult for many digital marketing teams to determine this marketing attribution. The problem is often due to the independence of the different components of platforms for cross-channel campaigns, which results in data being situated in a number of unrelated systems, making it challenging for teams to track ROI confidently. Determining the *attribution model* and how much weight to assign to which channel or touch point in the customer journey is a big challenge in the absence of adequate data. Undervaluing one source in the attribution model and overvaluing another can lead to bad spending decisions and poor marketing outcomes. Using omni-channel customer journey data, many companies use AI to create more accurate models to attribute results to specific channels.

Cross-Selling, Up-Selling, and Loyalty

For prospects who have already converted, there is more usable data available, which can be combined with third-party data, taking privacy and regulations into consideration. Insights gained from when that user was a prospect are copied over to the customer profile once she has become a customer. This data is then gathered in a data lake called a *customer data platform*, from which AI models are developed.

In the past, targeting customers was not granular enough for high-accuracy predictions. At best, it was based on coarse-grained customer segmentation. Traditionally, customer segmentation focused on dividing the customer base into groups that had specific kinds of commonality that were relevant for targeted marketing, such as age, gender, and location. Now, with the surge of available data such as browsing histories, ad impressions, and social media, companies are using AI algorithms to create fine-grained *customer segmentation* that is much more customer specific. This segmentation is then used as input in other models for greater accuracy of personalization.

Customer lifetime value modeling is another way to group customers, based on the amount of revenue or profit a customer will generate for the company over the course of the relationship. These models leverage machine learning to predict future purchases based on historical purchases, helping to prioritize recommendations and offers. This enables businesses to convert the highest-value customers: those who will be loyal and buy more through cross-selling and up-selling in the future.

E-commerce and retail businesses use AI models to predict customer behaviors such as purchase propensity: scoring people by the factors that are most predictive of purchasing within a specific period. These AI models for predicting customer behavior are tailored to specific industries and then segmented based on such things as behavior within that industry, customer buying power, share of wallet, and revenue. The resulting information is used for personalization, cross-sell and up-sell targeting, and calls to action, thus increasing sales conversion. Models such as *propensity to buy* and its underlying data are also being extended to improve other areas of businesses, enabling not only optimization of advertising spending on customers who are most likely to make purchases, which improves ROI, but also helping to improve demand forecasts in supply chain planning.

One approach to increasing customer engagement is using *site personalization* to provide each customer with a unique version of a website or app. Many elements of the site are customized, such as the banner image or video, the colors, content being displayed, and product or service recommendations; algorithms called *recommendation engines* provide the personalized content shown or product offered to each user. Most recommendation engines use a class of algorithmic techniques called *collaborative filtering*. These analyze data based on categories that focus on similarities both among potential customers and among products. Individuals are considered similar if they, say, purchase many of the same products, or share multiple characteristics, such as demographics, interests, and shopping history. Products are categorized as similar if shoppers tend to purchase them together, such as, say, mops and floor cleaners.

The AI model looks for what aspects of a user's extended profile (including transactions, browsing behaviors, and any insights from third-party data) match with what kinds of content she spends time on. In scenarios in which there may not be historical data, such as new individualized designs for the website, *A/B testing* is used. A/B testing, or more generally multivariate testing, is a way to compare alternative variations of a site (or alternate versions of AI models) by showing each variation to different visitors at random to determine which performs better given a particular goal, such as conversion.

Although collaborative filtering using deep learning has improved the accuracy of predictions, more modern AI algorithms have emerged, such as the wide-and-deep-model, which are much more promising in terms of accuracy. This neural network model combines an understanding of user and product interactions (wide network) and a rich understanding of the characteristics of products (deep network). Using AI-based recommendation engines, companies are predicting what products a customer will be most amenable to purchasing – and recommending these products through online channels or direct messaging. With the right algorithm, it is not difficult for, say, a clothing retailer to determine which products a customer might be interested in by looking at the similarities between this customer and previous customers who have purchased similar items.

Improving cross-sell and up-sell opportunities have huge potential upsides. Using a variety of AI techniques that utilize customer intent predictions and recommendation engines, retailers can create the optimal next-most-likely step in the sales process to increase sales of the additional products.

Many companies are using recommendation engines to create a continually improving virtuous cycle through repetitive experimentation and exploration. Quality data yields insights into how users notice or ignore products or follow recommendations; companies utilize these insights to continuously improve this virtuous cycle. Resulting decisions may include whether a company wants to price or bundle items differently or improve or eliminate certain features from a product entirely. The results of these decisions enter the cycle as data and in turn yield new insight, and so on.

AI algorithms are also helping companies predict and reduce *customer churn*. They identify unhappy customers early, giving the company a chance to offer incentives to encourage them to stay. The incentives may include upgrades, free features, or discounts on future months of service. We will use customer churn as a detailed example in Chapter 13, looking at historical records of customers and constructing and training machine learning models to predict the behavior of new customers. A close use case to customer churn is *replenishment models*. These models predict what products a customer had bought that they may be running out of and when. Before they are predicted to run out, they are sent a reminder to easily replenish that product so the purchase is not made with an alternate product or retailer.

Business-to-Business Customer Intelligence

Business-to-consumer (B2C) companies are not the only ones that are adopting AI to enhance performance; business-to-business (B2B) companies, from those selling products to companies to financial services and energy companies, are getting into the act. Historically, B2B salespeople relied on a primarily intuition-based approach using limited data (often on spreadsheets), tracking their insights manually in a time-intensive process. But many B2Bs have digitized their sales using cloud solutions from SalesForce and Microsoft.

Like a physical retail visit, every time a potential buyer visits a seller's digital property, that seller acquires information about her.

This information includes where she comes from and details of her company, such as name and size. The company that employs the visitor is identified using a reverse lookup of her IP address. The seller can also track what a visitor is doing on the site by collecting information about the time she spent on which pages of the site and what she did while she was there, providing deeper insights into her interests and intent. For example, if she spends most of her time looking at product or service comparisons, she is probably further along in her research and may be more likely to purchase. Even how far she scrolls down a page says something about where she is in her research and buying journey.

All this information is collected for each person from that particular company who visits the seller's website. If the company is an existing B2B customer, additional information about the company will be used: information such as past purchases, what the conditions of its usage were, the services and products used, how users interacted with a mobile app if one is available, and the outcome of their previous visits. B2B companies aggregate this kind of customer intelligence by company, enabling the seller to have a holistic view of the behaviors of all the users from a given business. The resulting data provides insights into both the business environment and user intent, allowing personalization of the site so the users' experience is even more relevant.

B2B companies also use other methods to collect data about their users' companies, using web scraping from news sites and job postings and from data purchases to understand more about them. Insights gained from this structured or unstructured data may include hiring spurts, investments made, changes in the board, or new strategic directions. Combined effectively, the resulting data creates a view into the company's "behaviors." Based on this information, the seller uses AI models to score the likelihood that users within that company will buy the seller's products and services, and if so, when. This insight then allows relationship managers to focus on the likeliest segments and buyers of the target accounts.

Not only can an AI algorithm score a company's best leads, it can continue to refine the score as the buyer comes into contact with the sales and marketing divisions. Algorithms are some-times even used to match a particular sales rep from the seller with a specific opportunity for the best results. Using AI mod-els, account managers gain new insights into customer activity, allowing them to quickly identify the best ways to customize their product or service or the messaging about that service to increase customer success. Professional services companies, industrial product sales companies, energy companies, and others use this type of approach.

Dynamic Pricing and Supply Chain Optimization

Port strikes or labor shortages in buyer locations are some of the third-party data that might go into an AI system used to predict demand. Companies use AI to leverage contextual data and build AI models that improve predictions. Information such as day of the week or week of the year, customer browsing behavior, and current marketing campaigns can all influence demand, and poten-tially, price.

The weather, for example, has a dramatic effect on how people buy or what they buy. Bad weather is good for e-commerce. People stay indoors, browse, and buy online rather than venture out to a store. Calculating this information ahead of time and combining it with other data points tells a retailer whether it will sell more or less during that day. Weather conditions can also affect the supply chain, which can have an impact on such things as warehousing and running low on inventory. Improving *demand forecasting* with artificial intelligence to enhance retail supply chains has shown sig-nificant benefits, enabling companies to make accurate forecasts of what products they need during which seasons, at what time of day or year. This improves the retailer's ability to keep product in stock, saving on inventory storage and even spoilage.

Using AI models for *dynamic pricing* also brings in price-sensitive customers without surrendering revenue from customers who are less sensitive to price. Having these data points means a business can price things differently when it is desirable to do so. The advent of flash sales is a direct result of this kind of knowledge, albeit at a more aggregate level. AI algorithms can set optimal prices in close to real time to improve revenue and profits for businesses and avoid the need for flash sales. This is easier for an e-commerce business, since all parts of the transaction are already digitized; but services are emerging in physical retail companies as well, where dynamic pricing labels on shelves in the form of mini-screens are enabling similar processes. These price labels are not targeted at individuals but are more based on time and place, and again, even weather.

AI models also help with *product returns*, a scourge of many businesses. In the United States alone, Statista estimates return deliveries will cost $550 billion annually by 2020.[4] Return rates are worse for online shopping compared to in-store shopping, but according to various surveys, free shipping and free returns are some of the critical reasons that buyers are likely to shop online. To deal with this dichotomy, many retailers are using machine learning models to discover the root causes for returns, such as poor fit, unexpected finish, or because of the "dressing-room" effect. The latter refers to a situation in which customers buy things as if they were trying them on in the dressing room of a bricks-and-mortar store, intending not to make a purchase until they see how the garments look on them. In e-commerce, this means buying many items and then, after trying them on, returning most or all of them.

Companies are also building AI models that evaluate an individual's shopping cart based on previous behavior – say, consistently ordering the wrong size – to determine how much they are at risk of returning the products. They then institute punishments or incentives if returns are likely. These include deterrents, such as increasing shipping charges, or incentives, such as offering coupons in return for making purchases nonreturnable.

The combination of online and traditional physical retailing, known as omni-channel retailing, has led to another development: the use of the traditional store location as a place to fulfill or return orders. Retailers often use AI for *fulfillment optimization*, determining the most cost-effective way to handle an order: shipping either from the factory, a fulfillment center, or a local store. Combining the best shipping models with global inventory models, AI enables these retailers to replace static rules such as "ship from the closest warehouse" to, instead, dynamically optimize for profitability.

Some larger retailers are experimenting with, or using, robots and drones in their warehouses. They often partner with robotics companies and startups for this work. Although most robotics uses AI and machine learning to operate the robot, this is a vast topic on its own and is beyond the scope of this book.

Digital Assistants and Customer Engagement

Customer service interactions have evolved over time. Initially, all interactions were over the phone with a live customer service representative (CSR). Then, businesses deployed interactive voice response (IVR) systems: those sometimes-annoying interactive voice prompts that direct you to respond to a variety of questions before taking action on your request or forwarding you to some-one who could help you ("To speak to an operator, press zero"). These worked via dual-tone multifrequency signaling (DTMF), the technology that allows you to respond via your phone's touchtone keypad, and they took some of the burdens away from the CSRs. Later messaging was added to the mix: a way to replace phone calls by chatting with CSRs. This enabled companies to compile conver-sations and handle problems more efficiently than phone voice-prompts alone could. The asynchronous nature of the channel saved companies time, since the pace of a conversation could be handled without the customer feeling like she was on hold, even though the respondent was able to handle several customers at the same time.

Now, messaging or chatting is increasingly happening between customers and AI-driven *digital assistants*, sometimes called vir-tual assistants or chatbots. AI enables *natural language processing* (NLP) and speech to text as well as text to speech translation within these digital assistants. This allows the assistant to understand customer questions and find the appropriate response or answer.

Juniper Research found that AI-enabled chatbots alone could save businesses $8 billion a year by 2022, with healthcare and banking benefiting most.[5]

Natural language processing is the ability of a system to tease out meaning from language. This ability to understand written (or spoken and transcribed) natural language enables these AI models to pinpoint specific information in large text documents, as well as filter and group this information, allowing companies to automate their responses to customer requests. The current state of NLP "understands" customer "asks" and answers questions when reasoning is not required – where the answer is written down in some unstructured text data that needs to be found and validated as a possible answer. Most customer-facing AI-enabled digital assistants today are integrated with agent-escalation, such as pressing zero to talk to an operator. This is because the AI-driven NLP technology has not matured to the point where all customer interactions are managed autonomously.

Leveraging NLP, AI is being used to examine historical call transcripts at call centers. The result is an enhanced ability to understand not just the frequency and timing of calls, but also caller sentiment and types and topics of requests: if, for example, seasonality affects the number of calls about a particular subject or the duration for specific requests. This in-depth knowledge enables AI algorithms to forecast more precisely the number of calls expected per hour per topic. The resulting improvement in matching and scheduling call center agents means contact centers are less likely to be understaffed, which negatively impacts customer service, or overstaffed, which is unnecessarily costly.

The use of NLP also enables contact centers to direct calls in a more efficient and even compassionate manner by freeing employees to handle more sensitive situations. Humans do not always have to handle requests; digital assistants may be appropriate if a caller will not feel alienated in the process. However, if someone calls to make purchases or update an account connected to, say, a bereavement, AI-enhanced call centers are much more likely to have trained call center agents available to handle the situation. This approach moves the high volume of simpler interactions from human agents to the digital assistant, leaving the agents more time to manage complex customer needs. Some companies are using these call center interactions and transcripts to continually feed the organization's knowledge graph, improving the future conversations that NLP can support. (See Chapter 12 for more on this pattern.)

With the help of AI systems, retailers are currently optimizing which channels are best for which type of message (phone, text, email, web, or app). AI systems are enabling retailers to gauge each customer's responsiveness to the channel as well as to the message itself, including to the length of the message and how it is worded. Even a relatively small change in how retailers reach their customers means customers may be more likely to see and respond to these messages. Such changes can additionally lead to substantial increases in a retailer's marketing-campaign success rate.

Digital assistants have a variety of uses in both the retail arena and many other industries as well, uses that may be relatively simple but are increasingly valuable to both these businesses and the customers they serve.[6] Available 24/7, these digital assistants are set up to handle many types of customer requests. More advanced digital assistants guide customers through a given process, such as a complex purchase of an appliance or a car. In financial services,[7] for example, they are used to walk customers through the steps necessary to apply for a mortgage loan, taking what would have been a painstaking process of filling out a lengthy application on a website and turning it into smaller conversational interactions.

These interactions help increase customer conversion, since customers who are uncertain or have unanswered questions are more likely to drop out of the purchase process. A digital-assistant-supported purchase more readily addresses and answers many of the questions that customers have. Customers who are already familiar with, say, Facebook Messenger, find this approach both convenient and time-saving. These voice or chat-based experiences are made possible by advances in machine learning and natural language processing, and their availability means that businesses are augmenting expensive human employee interactions with automated solutions.

Retailers are also taking advantage of the increasing adoption of smart home devices such as Microsoft Cortana, Apple Siri, Amazon Echo, or Google Home. They are integrating these intelligent devices into their AI systems to support sophisticated automated customer conversations, with a company-specific "skill" being added to the base device capability. Smart home devices work similarly to opening an app on a smartphone, but instead of clicking an icon to open the app, you say "Alexa, open my ENTER_YOUR_BRAND_HERE app." Once the app is active within Alexa (e.g. as an Alexa skill), the customer interacts with the retailer's app to authenticate, ask questions, and transact. Although retailers must consider what products work well in a voice-only environment, many are focusing on repurchases.

Currently, research is being done on ever-more-complex scenarios, such as enabling AI to detect, directly from the customer's tone of voice or use of language, when she is unhappy and likely to leave the business, although businesses have not yet deployed this capability at scale. This relatively new field, known as *affective computing*, involves using AI in the form of speech analysis, and facial expression recognition to understand and gauge emotional responses to a variety of stimuli by assigning values to positive, negative, or neutral texts.[8] Affective computing augments sentiment analysis, a type of text mining that uses NLP to determine people's opinions, enabling AI systems to analyze customer interactions, for example, to detect situations in which clients are expressing unhappiness.[9]

Many companies are now also using AI technologies such as machine learning and NLP to support employees to appropriately responding to both text and voice customer-service inquiries. This is often done by deploying the digital assistant to CSRs, rather than to the customer directly. When a customer calls in, a human CSR accesses a digital assistant that follows the conversation. In this way, the digital assistant is there to support the agent to help the customer, but the customer does not interact directly with the digital assistant. The digital assistant suggests to the human agent the next best action for this customer, and the agent relays this information to the customer in the most appropriate way. Some firms have also developed product and service recommendation algorithms in these digital assistants to help with cross-selling and up-selling. User transactions and navigation patterns are processed to generate recommendations and enable agents to offer customers the best products or deals available. They are likely to be the same recommendations the customer would receive if she or he were on a personalized mobile app.

These non-customer-facing AI technologies, which enable improved customer service, tend to be used more because using them internally means they can be tested and refined without the risk of alienating customers. This prevents inappropriate recommendations from going directly to a customer without human judgment and maintains empathetic customer interaction through a human agent.

Support centers often get emails from customers with headers that just say "Help" in their subject lines. NLP is currently the best-automated way to turn these emails into something actionable by digging more deeply into the text, knowing who in the company deals with a given customer or a problem, and automatically *routing problem ticket* to that person. Another current NLP application is to understand customer feedback or comments to glean insights that enable companies to make changes and improve processes.

One of the more interesting uses of AI that some retailers are experimenting with are "next-generation retail experiences." These include voice and gesture inputs integrated with touch screens or augmented or virtual reality (AR/VR) at retail kiosks and other experimental showrooms.

From precisely targeted marketing to individually personalized recommendations, the AI solutions described earlier are already in use in omni-channel e-commerce within retail and other industries. Many more AI applications are anticipated in the future.

Sidebar: Natural Conversations

One of the holy grails of AI is to create an algorithm that allows customers to have natural conversations with computers over the telephone, rather than relying on stilted IVR interactions or expensive human agents, which are the current norm. Natural conversations mean better use of time and higher profits without lowering satisfaction for the customer. In a move toward this ideal, on May 8, 2018, Google announced Google Duplex, a technology that allows people to carry on natural conversations over the telephone with computers.[10] Currently, the technology is very limited, directed toward particular tasks like making an appointment at the hairdresser. Even restricting Duplex to this kind of closed domain still requires that it be deeply trained. However, it is one step toward being able to have natural interactions with a computer.

Notes

1. *The Washington Post* (February 20, 2019). Digital Advertising to Surpass Print and TV for the First Time, Report Says. https://www.washingtonpost.com/technology/2019/02/20/digital-advertising-surpass-print-tv-first-time-report-says.

2. *Digital Commerce 360* (February 28, 2019). US Ecommerce Sales Grow 15% in 2018. https://www.digitalcommerce360.com/article/us-ecommerce-sales/ (accessed September 26, 2019).

3. *New York Times* (August 23, 2019). I Visited 47 Sites. Hundreds of Trackers Followed Me. https://www.nytimes.com/interactive/2019/08/23/opinion/data-internet-privacy-tracking.html (accessed September 26, 2019).

4. *Statista* (June 18, 2018). Costs of Return Deliveries in the United States from 2016 to 2020 (in billion U.S. dollars). https://www.statista.com/statistics/871365/reverse-logistics-cost-united-states/ (accessed September 26, 2019).

5. *Juniper Research* (July 24, 2017). Chatbot Conversations to Deliver \$8 Billion in Cost Savings by 2022. https://www.juniperresearch.com/analystxpress/july-2017/chatbot-conversations-to-deliver-8bn-cost-saving (accessed September 26, 2019).

6. *Emerj* (September 24, 2019). AI in Banking – An Analysis of America's 7 Top Banks. https://emerj.com/ai-sector-overviews/ai-in-banking-analysis/ (accessed September 26, 2019).

7. *American Banker* (October 24, 2016). B of A Debuts Virtual Assistant at Money2020. https://www.americanbanker.com/news/b-of-a-debuts-virtual-assistant-at-money2020 (accessed September 26, 2019).

8. PAT Research. What Is Affecting Computing. https://www.predictiveanalyticstoday.com/what-is-affective-computing/ (accessed September 26, 2019).

9. Digitalist Mag (October 18, 2017). Machine Learning With Heart: How Sentiment Analysis Can Help Your Customers. http://www.digitalistmag.com/customer-experience/2017/10/18/machine-learning-sentiment-analysis-can-help-customers-05429002 (accessed September 26, 2019).

10. Google AI Blog (May 8, 2018). Google Duplex: An AI System for Accomplishing Real-World Tasks Over the Phone. https://ai.googleblog.com/2018/05/duplex-ai-system-for-natural-conversation.html (accessed September 26, 2019).

Chapter 4
AI in Financial Services

We're witnessing the creative destruction of financial services, rear-ranging itself around the consumer. Who does this in the most relevant, exciting way using data and digital, wins!
Arvind Sankaran, partner at Jungle Ventures

Although banks, asset managers, and other financial services companies tend to agree that AI is essential, they vary widely in why, how, and when to adopt this new technology. According to an article in the *McKinsey Quarterly*,[1] more than a dozen European banks are now using machine learning techniques instead of older statistical-modeling approaches, and some are now seeing up to 10% increases in new-product sales; 20% savings in capital expenditures; and a decline in customer churn of 20%. In 2018, the World Economic Forum reported that "a growing number of financial institutions are applying AI to customer advice and interactions, laying the ground-work for self-driving finance."[2] One European bank, according to a McKinsey survey, was quoted in the *Financial Times* as having 500–800 people working in AI.[3] Although these metrics seem impressive, it is likely that many of the examples are smaller proofs of concept rather than enterprise-wide applications of machine learning.

Financial institutions have large volumes of structured data, often already of high quality, which makes it amenable to AI use. Many banks are using AI for assessing and managing risk, such as in fraud detection and managing credit risk to approve loans; for customer services, such as digital assistants and customized and predictive financial advice; for generating higher alpha with algorithmic trading; and for improving operational efficiency within banks.

As AI continues to prove its efficacy in banking, as well as in an increasing number of other industries, investment firms are already using it to make trading decisions, and they are willing to bet massive amounts of money on it – both their clients' money and their own. Some AI use cases are more relevant for companies looking for long-term growth, such as private equity firms; some use cases may be relevant for returns in the medium term, such as asset and wealth management companies. Still other cases may be most appropriate for quantitative hedge funds that are looking for a daily turnaround when entering and exiting an equity position (buying and selling on the same day or within milliseconds).

Two significant AI use cases in the investment banking industry are in the areas of investment research and algorithmic trading. The third primary use is within the sales and distribution organizations of asset management companies so they can understand client behaviors and better target services to meet their clients' needs, such as creating impactful personalized insights and experiences for clients and providing human relationship managers or advisers with the relevant information to help investors.

These use cases are just the tip of the iceberg in terms of the applications of AI in banks or asset management companies, and in the future, financial services institutions will increase the range of their use cases further. The following are some specific areas in which AI models are achieving successful outcomes.

Anti-Money Laundering

AI is becoming instrumental in dealing with one of the more pervasive problems facing the banking industry: that of money laundering. According to the United Nations Office on Drugs and Crime (UNODC), "The estimated amount of money laundered globally in one year is 2%–5% of global GDP, or $800 billion–$2 trillion in current US dollars."[4] Additionally, there are a variety of ancillary costs of money laundering, including such criminal activities as tax evasion, drug dealing, human trafficking, and the financing of terrorist activities.

Money laundering is the attempt to hide funds that have been obtained illegally by passing them through the financial system, primarily to conceal or disguise three things: where the funds came from, who owns or controls them, and where they are finally being held or spent. Money launderers deposit funds that appear to be legitimate but are derived from criminal activities into a bank or brokerage account or other financial institutions. Next, they perform a series of transactions to hide that first transaction. For example, money launderers might withdraw their funds from the first institution and deposit them into one or more businesses multiple times. After this, they might extract the money and spend or invest it.

All banks have anti-money-laundering (AML) teams on board to fight financial crime. They monitor customers' activities that may be indicative of money laundering. These teams are responsible for supplying regulatory institutions (which include governments or international agencies) with suspicious activity reports (SARs). In the United States, the Financial Crimes Enforcement Network (FinCEN) unit of the US Department of the Treasury regulates the banking industry. Suspicious activity reports are necessary whenever a questionable transaction is registered.

Dubious transactions that may indicate potential fraud are most easily detected by separating the wheat from the chaff: that is, distinguishing ordinary, expected transactions from criminal ones. To do so, AML teams use what is known as anomaly detection. First, they connect the dots among diverse sets of data from a variety of sources, likely housed in a host of disparate systems. This data might include information about previous SARs or existing AML data. Then they look for anomalies in the transactions.

Until recently, banks have been able to flag a variety of simple discrepancies, for example, a customer transaction in which deposited amounts were more than twice the normal level, which could be consistent with money laundering. When a matching pattern is observed, either one that resembled previous money laundering or a continued repetition of the same unusual transaction, an alert is generated and the case referred to the bank's AML team for manual review. After potentially multiple rounds of evaluations, if investigators conclude the behavior is indicative of money laundering, the bank files a SAR with its regulating agency.

The difficulty is that although banks have been using computers to monitor transactions for decades, these computers utilized rules-based systems. Any rules-based system is rigid by definition and therefore will not necessarily be able to parse the complex interactions among various ways used to launder money. That may be why the false-positive rates generated by such rule-based systems are as high as 90% or more. This makes their ability to successfully track down all real fraud cost-prohibitive, because the flagged transactions require many expensive person-hours to investigate.

AI, on the other hand, uses available data to create a collection of AI models. The AI engine then selects the models that best predict which cases in the data resemble those that ultimately resulted in past SAR filings, thus defining more accurate patterns before passing the information into human hands. This reduces the false positives significantly, while picking up more true positives. Generally, an AI system of this type works by initially applying unsupervised learning to the data, enabling banks to discover groupings by patterns as well as potential anomalies or outliers. After that, if there are known money-laundering cases, those data points are highlighted in each cluster, leading to a review of the group of data near which the fraud cases fall. As additional examples are found, either through this process or by discovering them through other investigations, enough data is collected to utilize supervised learning, which predicts whether new transactions are likely to be fraudulent.

Many banks are already using AI to flag unusual activities, enabling them to determine when an analyst needs to further investigate a transaction. This use of AI gives AML teams an excellent way to assess compliance with AML regulations, also reducing the false-positive rate for cases selected for manual review. The use of AI thus leads to more efficient compliance teams. Another advantage of an AI system is the ability to evolve as the behavior of money launderers changes over time, meaning that AML teams do not have to revise and re-implement rules explicitly. Still, it is crucial to recognize that new types of fraud or anomalies are continually emerging, so unsupervised learning will continue to be used.

Loans and Credit Risk

With rising expectations from commercial and retail customers and the potential to increase revenues while lowering risk, banks are starting to leverage AI for their loan approval decisions. Credit approval has been based on rules-based logic using static factors such as the business sector and turnover for the industry (for business loans), personal and business credit rating, marital status, and other factors. This approach has been the foundation of credit life-cycle management for approving credit cards, personal and business loans, mortgages, and other lines of credit for some time.

These factors tend to evolve slowly even as market conditions change, though consumer or business credit health can deteriorate quite rapidly. The key objectives for using AI are to increase automatic approval rates for lower transaction costs, improve time to decision, and enhance the quality of decisions with reduced risk and bias. To accomplish this, AI leverages massive datasets for the decision-making process, such as customers' transactional behavior through payroll, rent, and utility payments in addition to static profile information and traditional credit factors such as debt-to-income ratios, to improve loan decision quality. This data is then used to build AI models to predict credit worthiness.

The approach that some banks take first categorizes the transactions into relevant groupings. This categorization gives the bank a way to understand in what kinds of transactions each customer is engaged. Next, they look at time series analysis, for example, searching for predictors of financial distress or other behavioral risks such as making monthly loan payments with their credit cards. After that, they cluster the customers by credit default risk – now based on these inputs rather than solely on profile or static data such as external credit ratings, thus improving the quality of credit decisions.

Predictive Services and Advice

The depth of customer engagement has always been a central component in increasing long-term value for banking customers, and AI systems are used to improve both customer engagement and satisfaction. Previously, banks would call their top customers with advice or offers, and the sales and customer success teams recorded their insights for future discussions. This was only possible with a small percentage of valuable customers. Now, AI systems are helping to create customized engagement for all customers, based on tracking each customer's communications and interactions with the bank and the insights extracted from their transactional information and data about their interactions. The combination of AI's ability to handle massive amounts of data and capacity for executing sophisticated models is allowing banks to analyze a given customer's spending patterns and transactions relative to her situation and relevant global and local financial trends and incidents. These sophisticated analytics, driven by AI, help banks determine the best short- or long-term strategy for any given customer.

AI systems are using behavioral data to determine the state of that customer. This data comes from a variety of sources, such as bank and credit card transactions, the browsing history of the customer on the bank website or apps, contacts with the call center or branches, or other interactions with the bank. This information enables financial institutions to obtain real-time, substantive insights into what customers' needs, interests, goals, and concerns may be. These insights allow them to provide customers and clients with customized, end-to-end experiences, improving everything from providing financial advice to determining the best method of communication for any given customer, whether it be by phone, email, text, or other messaging such as WhatsApp or Facebook.[5]

Banks are taking their cue from e-commerce recommendation engines in the retail industry to offer their customers predictive financial advice that is suited to both their financial needs and their emotional comfort. The analyses enable the bank to provide each client with recommendations specifically tailored to her, creating more personal and productive relationships. AI application-based services such as these are poised to become increasingly important to a bank's profitability. This transformation is evolving along the lines of the personalization transformation in retail, where there is a big difference between recommending the most popular item and making a personalized recommendation.

One example is the use of AI to predict future transactions for customers, based on a variety of customer data, including such things as historical spending information from credit card bills and bank transaction information. Predicted future transactions are used to forecast spending and account balances, and based on this information, the bank makes recommendations about actions the customer can take to avoid an overdraft. Avoiding overdrafts can be a significant benefit to the customer – in 2018, US bank customers paid overdraft fees of $35 billion. Conversely, if the customer is predicted to have more money than usual, the system may recommend transferring money into savings or investment accounts or to make extra payments on loans, if these are a better option.

Another example of an AI application is to predict future customer behavior, such as an appliance or car purchase. These predictions may be based on recent transactions such as car repair payments on the bank's debit card or clicking on the bank's digital car-loan ad on a car website. In these cases, credit is preapproved in advance, assuming the customer qualifies, and the customer receives messages in anticipation of the purchase. Other examples include predicting income taxes based on annual income and expenditures, or recommending a mortgage refinance based on interest rate changes. Because bank transaction data is such a rich data set, many kinds of predictions are possible, and different banks are taking different approaches.

The same types of AI used by financial advisers are also being applied to various use cases within wealth management. In 2017, Morgan Stanley's wealth management division brought in almost half the company's annual revenue.[6] Augmenting a wealth management group with AI could dramatically increase a company's profitability. Because relationship managers mediate much of wealth management, banks are taking a hybrid approach, offering customers a variety of AI-enabled client-facing tools, such as self-service onboarding, portals, and interactive dashboards. Relationship managers use these AI-enabled tools to get recommendations for the next best action to take with their top set of clients. Infusing AI within the wealth management process is unlocking new levels of client-centricity, organizational efficiency, and cost savings.

As financial services companies roll out cognitive wealth management algorithms, front-office functions are changing in two principal ways. Relationship managers and the wealth management team now use their time more efficiently and effectively, and clients have the option of getting human or virtual support. By using AI-driven automation rather than costly human-to-human interactions, banks also increase profit by making it cost-effective to serve larger populations of less affluent clients, bringing a whole new set of customers into the organization.

Sidebar: The Black Box Problem in Financial Use Cases

Financial companies trying to understand exactly how AI algorithms make the decisions they do must contend with the challenges inherent in the technology. Chief among them is the fact that AI decisions occur within a black box. In this situation, you know what data you put into the system and what predictions come out of it. However, it can be difficult or almost impossible to figure out how the system made the prediction it did, depending on the complexity of the models. This lack of transparency can become a serious problem when businesses must explain why their algorithms messed up and lost money for their customers. It is something to remember when looking for AI use cases in any industry. However, despite skeptics' beliefs that only human-level AI will be able to handle the enormous amount of information that affects financial markets, AI scientists continue to change the way financial institutions are doing business by addressing these challenges with interpretable AI.

Algorithmic and Autonomous Trading

Algorithmic trading has been around in different forms for many decades, sometimes under the names of automated trading and high-frequency trading. American financial markets began to use it in the 1970s. Algorithmic trading does not have to use AI; for example, a trader could decide to sell when the average of the last three ticks (of a stock price) was higher than some number, such as a historical moving average. This is an example of an algorithm or trading rule that leverages domain expertise (although this specific example is overly simplistic). Algorithmic trading quickly gained market share because it enabled trading organizations to make enormous profits. As AI algorithms matured, applying it to algorithmic trading was inevitable.

High-frequency trading, a form of algorithmic trading, had an execution time of several seconds when it was introduced in 1983. By 2010, this had decreased to milliseconds or less. As recently as 2016, capital market research firm Tabb Group estimated that high-frequency trading accounted for just under 50% of the average daily trading volume.[7] However, when profits declined as competition increased, traders began to look for something to give them a competitive advantage, and they actively started to use AI. Computer engineers developed ML algorithms that included information such as exposure caps, asset classes, and trading costs. When these algorithms were provided with enough data, they were able to compare it to historical patterns, resulting in insights that human analysts could not see.

AI was able to speed up the search for successful trading strategies, streamlining what had been, until then, a tedious and time-consuming process. It also increased the number of markets a trader could observe and act on. Finding associations manually from the data was now a thing of the past; AI allowed traders to identify and adapt to trends as they developed. This resulted in algorithmic trading becoming more lucrative. One reported use case is JP Morgan's AI program LOXM, which executes trades across its global equities business.[8] Unlike rules-based trading, LOXM learns from past trades to set its parameters rather than rely purely on trader input. It then executes orders at the fastest speeds and best possible prices.

Its growing intelligence enables it to tackle problems such as how best to offload big equity stakes without moving market prices. In the future, LOXM could get to know each customer even more intimately, enabling it to take his or her behavior and goals into account when deciding how best to trade for his or her portfolio.

Using AI in algorithmic trading means that AI learns what the rule should be, determining the structure of the data and making predictions based on that. Machine learning and deep learning are increasingly making these highly complex, high-speed transaction decisions. The primary approach for this has been predicting *time series*. A time series is a sequence of data points, such as a stock price, in chronological order and equally spaced in time. In algorithmic trading, AI is used to understand the structure of this data to forecast the time-series into the near future.

The difficulty of this task is that the data can seem random – and it is impossible to accurately forecast a random variable. However, computer scientists are successfully using many techniques to overcome these challenges. For example, they are employing feature engineering on the source data by aggregating by the number of events rather than purely chronologically, such as every time a particular volume of shares of a specific stock has traded, rather than by a time interval. In addition to this type of feature engineering, deep learning is coming into greater use within the quant community – that is, businesses that use systematic or algorithmic strategies – leading to what some are calling Wall Street's "Quant 2.0" phase.

AI-driven algorithmic trading is a focus of the quant hedge funds, where it is in the domain of high-frequency trading. Investment banks and asset managers have put a greater focus on using AI in the medium- to low-frequency trading domain to leverage publicly available information faster and at greater scale. That is, banks and asset managers are using AI to extract useful information from publicly available data before others act on these signals, and doing this across equities, industries, sectors, and regions (see next section). Using AI to make investment decisions involves research analysis for monthly or quarterly transactions. Unlike algorithmic trading, these are low-frequency autonomous transactions.

Investment Research and Market Insights

Active asset management is about the arbitrage of information and how quickly and inexpensively you can get this information to the decision makers or portfolio managers so they can act on it. Information management has always driven success in investment, from data collection to analysis and decisions. The critical thing is to gain the most relevant insights from information at the right speed and for the lowest cost. Investment research analysts spend a great deal of time on data gathering and organizing, but their productivity is becoming more and more critical as the financial industry becomes more competitive. Today, much of that time-consuming data management work is easily automated and analyzed using AI.

Financial data available to analysts used to be primarily of two kinds. One kind was in the form of information self-published by a given company in such vehicles as annual reports, SEC filings, press releases, and the like. The second kind was information coming from brokers or stock exchanges, such as stock prices, price history, and price-to-earnings ratios. Not only was this data the most readily available; other potentially useful information was in the form of unstructured data and hence difficult to use other than manually. *Unstructured data* is typically text-heavy, such as data from emails or tweets, although it might also contain elements like dates, numbers, and facts, as well as images and video. *Structured data*, on the other hand, is data in which there are well-defined fields, such as name, address, and phone number.

Before the availability of AI algorithms, unstructured data was far too time-consuming to utilize heavily. However, that is changing. AI algorithms, in the form of natural language processing, deals with large quantities of unstructured text data culled from a variety of sources. This includes data buried in filings, news, research, brokers, and in-house content as well as data from complicated publications with a variety of data elements, all of which have become essential to understanding how companies are operating. This is often called *alternative data* – nonfinancial datasets about a company that have been published by sources outside the company.

These datasets may come from a wide range of sources, including news feeds, job posts, social media discussions, satellite imagery, credit card transactions, and mobile devices. These are all unique, granular, time-stamped data that enables an extensive view of actual fine-grained trends. It often complements available financial data, helping analysts make better investment decisions by getting a broader base of inputs in their decisions.

AI models are also being used to mine data from organizations such as central banks and the National Bureau of Labor Statistics, compiling it into models that predict fixed-income performance, based on similar employment changes in the past. The inferences from these models are made within seconds, and resulting insights are applied within minutes of when the data is collected. AI systems enable firms to machine-read SEC filings, earnings calls, transcripts from investor days, and a variety of other sources. They also understand spoken or written natural language questions and provide answers, albeit only in the domains in which they are trained, thereby improving research analysts' productivity and the quality of their recommendations.

Once all this structured and unstructured data is available, natural language processing algorithms are used to "understand" the content semantically as well as to model an evolving ontology based on this understanding. When AI algorithms map out documents or even sentences within the content in this way (for example, as a knowledge graph), it is easier to pinpoint relevant information. For analysts who may otherwise spend much time looking for insights, this provides substantial benefits. One company did a study to see how much time their business users spend on which type of activity and discovered that over 36% of their time went to looking up information.[9] Leveraging AI algorithms and natural-language processing, data is organized so that analysts can search out information more quickly and precisely, enabling them to identify patterns and forecast performance more accurately and quickly. By layering all this data, investment analysts are not only saving time; they are acquiring more significant market insights. AI systems search millions of relevant documents and find not merely the one that best provides that information but a specific answer to a question that an analyst has asked.

Prior to the development of AI algorithms, research analysts within the asset management division of a company would read documents to answer a set of questions to enable them to decide whether to recommend buying, holding, or selling equities. They looked at questions such as "Is the company's revenue growing?" (from structured data), or "Is it in a strong-enough position to beat out its competitors in the future?" or "Does the management team have confidence in the company?" (from unstructured data). With the development of AI, AI scientists train natural language processing (NLP) algorithms on what to look for – for example, hedge words (e.g., "almost," "likely," etc.) used by management during earnings calls or analyst discussions can imply that they have low confidence in what they are saying. These NLP models are used to find answers to the analysts' questions. Then, the research analysts review a report that provides the questions, and the answers based on machine understanding, with links to the sources from where the answers were gathered. This allows the analysts to deep dive into the next layer of analysis if needed, and to cover more equities per day than before.

When evaluating industries to focus on, analysts are looking for answers to questions such as "Is the demand in this sector expected to grow in the near future?" or "Is industry disruption being discussed?" Both of these can be derived using unstructured text from news or analysis. For example, NLP is used to read multiple news sources and classify if the report is about supply or demand. If it is about demand, NLP then classifies the demand as increasing or decreasing per the news item. The output of this is a supply and demand sentiment time series, which is then used with historical industry growth indices as an input feature to predict future growth.

Traditionally, figuring out foot traffic for a business in a given area was done in two ways: by a tally counter, using a clicker to manually count everybody who walked by, and by setting up security cameras outside so employees could watch the footage on fast-forward and estimate the number of people who passed by per hour. Both were incredibly time-consuming, and the data was usually not available to investment companies. There is a scalable solution, however: using ad-impressions data from location-enabled mobile phones, which reveal more about their users (age and gender, for example) than a simple count. Financial analysts now have AI algorithms analyze satellite images of parking lots or mobile phone geolocation to keep tabs on how many people in the aggregate are visiting the stores of a specific company over time – with foot traffic or parking lot usage trends. Using this information over time, AI scientists predict future revenue, and having this prediction before the company's earnings calls gives the asset management company a leg up.

By aggregating all these datasets and overlaying them on existing financial models, financial analysts are improving their predictions of earnings for retail stores and other companies, as well as obtaining more in-depth insights into supply and demand and other economic factors. Equity research analysts, with help from AI models, are now automatically tracking stocks and developing time-sensitive short- and long-term perspectives on whether the followed stock will go up or down. Moreover, as raw data that influences stock prices keeps increasing, AI models are helping convert this unmanageable data into usable insights for portfolio managers.

All this data is not free. As with most valuable commodities, there is now a thriving market in alternative data (see Figure 7.1 in Chapter 7). Some hedge funds are spending millions of dollars to acquire it from vendors who source, clean, and sell it to the financial community. Many companies are also investing money to build their own infrastructure and creating their own sources of alternative data. Venture capitalists and private equity companies are also using AI to make investment decisions. According to the *Financial Times*, a partner in Stockholm's EQT Ventures, Andreas Thorstensson, makes around 30% of his investment decisions because of data analyzed by their AI platform, Motherbrain.[10] (EQT Ventures is the venture capital arm of the Swedish company EQT Partners.) Motherbrain daily monitors around two million companies, and as a result, Thorstensson is no longer investing in certain start ups. "The data doesn't lie," he has been quoted as saying.

Sidebar: When Faces Speak Louder Than Words

After the regular internal policy meetings held by the European Central Bank (ECP), the ECP president typically holds a press conference. The problem is, at that conference, he or she reveals little about future monetary policy. During the tenure of ECP president Mario Draghi, however, Japanese researchers were able to identify a correlation between patterns in his facial expressions and subsequent policy changes.[11] To do this, they used Microsoft's Emotion API, which employs a visual recognition algorithm to break down facial expressions of happiness, sadness, surprise, anger, fear, contempt, disgust, and neutrality. By analyzing Draghi's expressions, the researchers were hoping to determine the likelihood of policy changes slightly ahead of time, which would presumably give them an investment edge. The success of this analysis has not been determined yet, but it is an excellent example of how much AI is bringing to financial and other decisions.

Automated Business Operations

Robotic process automation (RPA) is the automation of high-volume processes for which people were formerly responsible. It is an evolving technology that combines keystrokes automation with business rules and AI to automate repetitive processes not requiring significant decision-making. Companies are currently combining RPA with AI technologies to help implement back-end process automation, processes that have up to now been mostly manual and thus were slower and less accurate, as well as less efficient. This "intelligent RPA" makes these processes quicker, more reliable, and more efficient, reducing human error and freeing up humans for other, more complex tasks. It also makes it easier to scale resources to meet changing demand by activating additional bots. Intelligent process automation (IPA) substantially augments the workforce with a team of software robots.

In financial services, there are many uses for IPA. For example, investment banks are automating several manual processes that ensure that any changes to instrument data arriving upstream are reflected in the downstream systems. IPA transforms this function, using robots to retrieve data exceptions from the source system and enter the information downstream. AI-powered RPA solutions are also currently aiding banks with the labor-intensive task of reconciling invoices and payments. Some banks are improving straight-through reconciliation of incoming payments by employing RPA, machine learning, and optical character recognition (OCR). This use of AI-powered RPA means the bank posts receivables more quickly, which makes reconciling accounts receivable and payment matching more efficient.

One bank has reported[12] that its RPA implementation of 220 bots resulted in significant process improvements, including 100% accuracy in account-closure validations, 88% reduction in processing time, and 66% improvement in trade entry turnaround. Other banking firms are implementing processes like these as they strive to remain competitive. According to reports,[13] IPA has helped another bank improve productivity and augment its workforce. Applications in trade finance, cash operations, loan operations, tax planning, customer onboarding, and others have achieved between 30% and 70% automation, improving quality and decreasing risk. IPA has also reduced the time it takes to train employees. By encoding knowledge through robotics, information becomes available to augment employees' skills and guide them through their everyday responsibilities.

Financial services and banking institutions are rife with opportunities for AI systems. In addition to implementing their solutions, such as the preceding examples, these organizations are also investing in, acquiring, or integrating services from the hundreds of FinTech companies that have emerged over the last decade that offer various types of AI-based services. The application of AI to financial services remains an area that will continue to see ongoing evolution, if not revolution, in the years to come.

Notes

1. McKinsey & Company (June 2015). An Executive's Guide to Machine Learning. https://www.mckinsey.com/industries/high-tech/our-insights/an-executives-guide-to-machine-learning (accessed September 26, 2019).

2. World Economic Forum (August 2018). The New Physics of Financial Services. http://www3.weforum.org/docs/WEF_New_Physics_of_Financial_Services.pdf (accessed September 26, 2019).

3. *Financial Times*. AI in Banking: The Reality Behind the Hype. https://www.ft.com/content/b497a134-2d21-11e8-a34a-7e7563b0b0f4 (accessed September 26, 2019).

4. MENA Report, Albawaba (London) Ltd. (March 2017). United States: Member States Must Strengthen Cooperation on Law Enforcement, Sharing Intelligence in Struggle against Human Trafficking, Secretary-General Tells Security Council. https://www.unodc.org/unodc/en/money-laundering/globalization.html (accessed September 26, 2019).

5. Wells Fargo (July 26, 2017). The Most Helpful "Banking Assistant" on Facebook. https://stories.wf.com/helpful-banking-assistanton-facebook/ (accessed September 26, 2019).

6. CNBC (July 19, 2017). Wall Street Banks Have Found a Great New Way to Make Big Money Now That Trading Is Down. https://www.cnbc.com/2017/07/19/forget-trading-wealth-management-the-real-driver-of-bank-returns.html (accessed September 26, 2019).

7. MarketWatch (March 17, 2017). High-frequency Trading Has Reshaped Wall Street in Its Image. https://www.marketwatch.com/story/high-frequency-trading-has-reshaped-wall-street-in-its-image-2017-03-15 (accessed September 26, 2019).

8. *Financial Times* (July 31, 2017). J.P. Morgan Develops Robot to Execute Trades. https://www.ft.com/content/16b8ffb6-7161-11e7-aca6-c6bd07df1a3c. *Financial Times*, July 31, 2017 (accessed September 26, 2019).

9. CMS Wire (February 22, 2017). Information Management: The Critical Thing You're Overlooking in the Digital Workplace. https://www.cmswire.com/information-management/information-management-the-critical-thing-youre-overlooking-in-the-digital-workplace/ (accessed September 26, 2019).

10. *Financial Times* (December 11, 2017). Artificial Intelligence Is Guiding Venture Capital to Start-ups. https://www.ft.com/content/dd7fa798-bfcd-11e7-823b-ed31693349d3 (accessed September 26, 2019).

11. Reuters (March 20, 2018). Japanese Researchers Seek to Unmask Draghi's Poker Face to Predict Policy Changes. https://www.reuters.com/article/us-ecb-draghi-face/japanese-researchers-seek-to-unmask-draghis-poker-face-to-predict-policy-changes-idUSKBN1GW14B (accessed September 26, 2019).

12. BNY Mellon Press (March 2017). The Rise of Robots. https://www.bnymellon.com/us/en/who-we-are/people-report/innovate/the-rise-of-robots.jsp (accessed December 5, 2017).

13. CIO (February 24, 2017). RPA Proving Its Transformational Value at Deutsche Bank. https://www.cio.com/article/3174126/rpa-proving-its-transformational-value-at-deutsche-bank.html (accessed September 26, 2019).

Chapter 5
AI in Manufacturing and Energy

Artificial intelligence is here and being rapidly commercialized, with new applications being created not just for manufacturing but also for energy, healthcare, oil, and gas. This will change how we all do business.

Joe Kaeser, CEO of Siemens AG

The use of data from the Internet of Things (IoT) and AI and machine learning within manufacturing has gained momentum due to the Industry 4.0 program in Germany[1] and the US Advanced Manufacturing Initiative in the United States.[2] The latter suggested launching an "Advanced Manufacturing Initiative" that would "support innovation in advanced manufacturing through applied research programs for promising new technologies, public-private partnerships around broadly-applicable and precompetitive technologies, the creation and dissemination of design methodologies for manufacturing, and shared technology infrastructure to support advances in existing manufacturing industries."[3]

Both the manufacturing and energy sectors are particularly well suited to the use of AI because they produce enormous amounts of data in a variety of formats, and they will continue to do so into the future. Much of this data is collected using sensors that monitor everything from the equipment used to the rate of a chemical process. Connecting these sensors to computers through wireless or wired networks enables AI algorithms to use this data to improve such things in the manufacturing process as quality, productivity, distribution, employee safety, and environmental impact.

Industrial manufacturers capture data through sensor readings such as voltage, pressure, temperature, or vibration of plant machinery. Other key parameters include information about the production run itself, such as production counts, machine availability, and quality. Additional complementary information may come from error logs, collected while machinery is operational and during maintenance events, both scheduled and unscheduled; from enterprise resource planning (ERP) systems; and from human operators or elsewhere. Applying AI models to this data, businesses identify the issues that increase production costs and predict situations that result in unplanned downtime. Knowing, for example, how each machine's load level affects overall performance enables better decisions during each production run. AI also helps determine which equipment is best for a given run of a particular duration, leading to further efficiencies not only in production but also in maintenance.

Energy companies similarly utilize sensors and wireless networks to collect data on everything from plant operations to individual consumption patterns to global weather events that lead to power outages. They then use AI models to leverage this data toward ways to improve productivity and cut costs. Algorithms parse the data and make predictions about how to handle unplanned events, assisting employees with such things as determining when machines need maintenance, monitoring personnel safety, and determining production efficiencies.

Both the manufacturing and energy industries store some of the data they collect as a time series in a centralized data store known as a *process historian*. Process historians may have data from as many as 10 or more years back, providing enough information to make the use of AI not only viable but also effective. Sometimes, much of this data cannot be accessed remotely. However, aggregating it in a data lake and uploading it to the cloud is now more common, enabling companies in a variety of industries to implement what is known as a *digital twin*.

Renowned advisor and professor Michael Grieves first used the phrase "digital twin" in 2002, when he was working at the University of Michigan. A digital twin is a virtual digital model of a process, product, or service that monitors, collects, and records data from the machines involved in physical processes in a manufacturing facility or energy plant and reproduces these processes virtually. In the strictest sense of how the term is used today, the digital twin is not the component controlling the process. It is usually thought of as the data model at the core of the digital representation of the physical equipment or process. Digital twins provide the context to the constant streams of data generated from sensors and other tools in production operations. Algorithms then utilize this context to make decisions and relay these back to control systems which then pass on instructions to physical actuators and control equipment. Over time digital twins will evolve from just the physical production process to a full-fledged view of the enterprise's value chain, covering their supply chain, production processes, finished goods inventory, distribution, marketing, and end customer delivery. Due to the introduction of AI and the IoT, implementing digital twins has not only become cost-effective, but also Gartner named them one of the Top 10 Strategic Technology Trends for 2017.[4]

After gathering and analyzing data, a digital twin discovers problems and finds solutions to improve procedures, lower operations costs, reduce product defects, and improve efficiencies, all with very little human intervention. Connected in a digital ecosystem of equipment, digital twins receive data through *sensors*, and control machines through what are known as *actuators*. They then communicate directly with physical machinery and control processes based on the conclusions they have reached. Digital twins are most often utilized to improve the performance of complex physical assets, such as refineries or manufacturing equipment, but there is a growing trend toward using them for consumer appliances such as connected dishwashers, refrigerators, and coffee makers.

One example is Whirlpool's Smart All-In-One Washer and Dryer,[5] which allows customers to remote-operate the wash and dry cycle and track progress using a mobile app.

Optimized Plant Operations and Assets Maintenance

To be competitive in cost-sensitive markets, getting the most out of expensive manufacturing machines is vital. Companies utilizing AI-controlled digital twins achieve process optimization in a variety of ways. Control algorithms have been around in manufacturing and energy for a few decades, with microprocessor-based programmable logic controllers (PLCs) coming into use in the 1980s. The use of AI helps to get an additional level of efficiency out of these industrial processes by leveraging the historical input and output data points from the control system, and the impact it had on the process, to optimize the parameters of the process. One crucial example is automatically controlling high-speed procedures, such as in discrete manufacturing, to achieve the maximum production output, the least amount of downtime, and greatest reductions in parts failure. Using real-time monitoring along with AI makes shop floor operations more efficient by improving production schedules and optimizing mechanical efficiencies. Managing production runs with AI helps determine how load levels impact production schedule performance in real-time, leading to more efficient production runs. These improvements together maximize equipment utilization. In addition, inventory and spare part optimization can also create substantial savings.

AI models help managers determine the best combination of machinery for any given run. Before production, the parameters for demand, raw materials, and other factors are given to the AI algorithm so it can determine optimal shop floor operations. During a production run, the equipment records and evaluates real-time data. Then, accounting for any discrepancies between the plan and actuals, the models are rerun and the execution plan updated to stay on target with the goals. The history of this physical production run is used to adjust the models as necessary.

Siemens, for example, has an AI system that utilizes data from over 500 sensors attached to its gas turbines so that it learns to continuously adjust fuel valves to create optimal combustion regardless of weather conditions or the state of the equipment. Additional benefits include fewer emissions and better pollution control. "Even after experts had done their best to optimize the turbine's nitrous oxide emissions," says Dr. Norbert Gaus, Head of Research in Digitalization and Automation at Siemens Corporate Technology, "our AI system was able to reduce emissions by an additional ten to fifteen percent."[6]

One of the biggest problems associated with manufacturing or energy production is unplanned downtime caused by mechanical issues. Companies now use AI models to reduce the frequency of this happening. AI applications are used to predict mechanical problems in advance so they can be dealt with before they begin to cause problems.

Hundreds or even thousands of smart sensors and the ability to collect data on a centralized platform enables AI algorithms to perform *predictive maintenance* on machinery, finding patterns of system degradation or failure based on historical data. These patterns result in businesses more optimally scheduling maintenance or replacement of machinery. This means, rather than resorting to emergency shutdowns, companies incorporate these actions into regularly scheduled downtime. When multiple parts of the machinery have predictive maintenance, the planned downtime can be optimized to handle multiple parts repairs or replacement. If this were unplanned, each part may have caused downtime when it failed. Even a small improvement in asset management has the potential to both save a business money and increase revenue. The US Department of Energy has indicated that in the power industry, for example, companies see up to a 70% reduction in breakdowns and a 35% reduction in downtime when companies initiate a functional predictive maintenance program.[7]

Consider an example. Producing low-density polyethylene (LDPE), a thermoplastic made from ethylene, requires an enormous amount of pressure. This pressure is provided by something called a hyper compressor, which may apply up to 50,000 pounds per square inch of pressure to form ethylene into this essential plastic grade. This amount of pressure puts a lot of strain on the hyper compressors, and they go down many times per year. A way to mitigate this problem is essential to LDPE manufacturing businesses, since the downtime can cost them millions of dollars per year.

Using AI models, companies automatically calibrate equipment based on a history of its performance as well as its settings, making an enormous difference in unplanned downtime. This type of disruption costs approximately $50 billion annually[8] across manufacturing, and estimates suggest that preventive maintenance spearheaded by AI models will reduce these costs by 10% to 20%. Some manufacturers have even used machine learning to not just predict when there will be a failure, but also identify when the component is at the end of its lifecycle and needs to be replaced. These AI models are built to be able to tell whether the component is close to its productive lifecycle based on failure rates and other information (see Figure 5.1). The data from patterns of use and equipment life expectancy can be used to create AI models that look to optimize a machine's life expectancy.

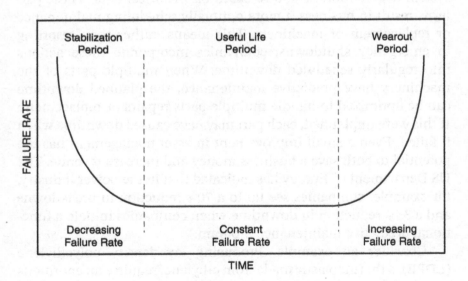

Figure 5.1 Heuristic showing different failure rates during equipment component lifecycle.

Automated Production Lifecycles

Production lifecycles that are either partly or completely automated are increasingly controlled and coordinated by AI models. Information systems, particularly in high-capacity manufacturing environments, plan, coordinate, control, and evaluate mass-production manufacturing processes that produce goods such as automobiles, dishwashers, and toasters. Having the flexibility of AI models allows for the adaptation of production lines to small-lot or customized production that would otherwise interfere with normal production runs. This enables manufacturers to more easily create personalized goods that consumers find increasingly desirable. Information systems also support workers on the shop floor, bridging the divide between the machinery and human workers. When workers are better informed about how processes work at given stations, the better they are able to make informed and appropriate decisions.

Quality control is another area in which AI algorithms are taking over for human beings. People, not machines, are usually responsible for quality control because it is either a visual function, or entails lab analysis. It used to be easier for a person to notice colors that are off, labels that are out of alignment, or packaging that is damaged. Deep learning-based image-processing models, however, now enable machines to inspect products and identify defects more quickly and accurately than humans can.

Supply Chain Optimization

Efficient supply chains have a significant impact on a company's success, and AI is well-positioned to make substantial improvements in demand forecasting, inventory optimization, and distribution logistics. AI is being used to *forecast demand* so that participants in the supply chain function optimally. McKinsey predicts that machine learning can reduce supply chain forecasting errors by 50% and reduce lost sales by 65% by making products more readily

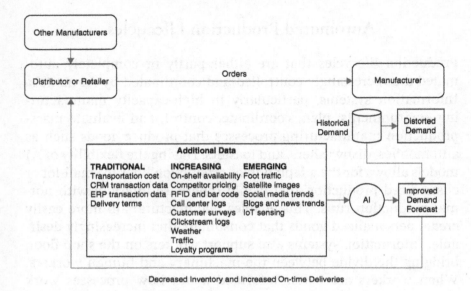

Figure 5.2   Demand forecasting using historical sales and new data sources.

available when needed. AI applications also help bring down costs of warehousing and transporting goods as well as those of supply chain management.[9] Some companies are even using AI to scale inventory optimization across distribution locations, using existing sales data to determine demand as well as time-to-customer delivery performance.

Increasingly, demand forecasting includes different types of data (see Figure 5.2). Manufacturers collect sales and inventory data across the network of retailers, gathering information about past sales at each location by product. Because most manufacturers do not often sell directly to consumers, they are incorporating marketing information in the forecast process. Using marketing information requires natural language processing (NLP) from sources such as news and blogs, product ratings, and other sources. Retailers have the advantage of incorporating weblogs and customer clickstream information, data from loyalty programs, call center transcripts, and customer surveys. NLP then extracts sentiment and demand trends into a time series, which can be used to reduce demand forecasting errors.

Manufacturers also collect supplier information, either by requiring suppliers to provide the information or gathering it from public sources. The supplier information often includes the supplier's current capacity and top customer size, as well as the supplier's prior transactions with other customers with similar requirements, such as delivery locations and lead times. Some companies are even experimenting with the inclusion of weather and traffic information, such as traffic congestion, in their AI modeling process to improve accuracy.

For situations in which there is no historical sales data, such as new-product introductions, companies use proxies in the modeling process. Forecasting demand for new offerings is critical, since these new products tend to drive high sales. To understand how much attention is being paid by potential customers to a variety of products, both new and old, AI models are given search query information: how many times someone has searched for a given term by geographic location and time. This information, together with other available data such as sales of related products, is used as an input in the ML model to predict demand

Inventory Management and Distribution Logistics

The goal of any manufacturer or distributor is to maximize sellthrough. The constraint for them is limited shelf space and often no storage space. Over-stocking one product can lead to decreasing potential sales of another product that either sold out or was never stocked at all. To address this problem, companies optimize their *inventory management* by enabling AI algorithms to use existing demand forecasts and other data to optimize stock levels and reduce loss by predicting circumstances that might affect product delivery.

Distribution logistics has been defined as getting the right product, in the right quantity, with the right quality, at the right place, at the right time, at the right cost for the right customer.[10] Products may need to be distributed from warehouses to various types of retail locations, such as electronics stores, supermarkets, gas stations, and convenience stores. Some retailers have the added challenge of shipping from warehouses or retail locations to the end customer. Historically, the results of operations research models based on data such as origin and destination (OND) were used to determine such things as vehicle routing, and utilizing logical and mathematical relationships to determine the most efficient way to ship goods. The ability of AI models to learn from experience provides a near-real-time alternative to these models.

AI models for demand forecasting, inventory optimization, and distribution logistics need to interact with each other because they are co-dependent. Data generated by marketing and sales is used to train AI algorithms, such as demand forecasting models, within the supply chain. In Chapter 3, we saw that the functions of advertising, marketing, sales, and customer service are starting to overlap more and more because they use similar models and the same data sets; the resulting insights are shared among these functions. As usage of AI increases within heavy industry and manufacturing and energy companies, there will be more overlap of those customer-facing functions with supply-chain functions there as well.

Electric Power Forecasting and Demand Response

If electric companies could better forecast demands for energy on a quarter-hourly basis, there would be less price volatility and fewer outages. However, this has been difficult to do because of subtle variations in demand and complexities within the power grid. Successful energy grid operations require the ability to continually redirect load to provide precisely the amount of power needed at any hour of the day to both commercial and residential customers. Predictions of the amount of energy required and to whom it should go must be highly accurate, which is challenging to begin with, and on-site power generation by solar arrays and on-site or grid-level storage, such as batteries, makes it even more difficult.

To respond to sudden spikes in demand, electric companies are required to provide additional power when requested by the grid operators, usually on short notice. Generators are able to provide this power because of "spinning reserve" – that is, the extra generating capacity available to generators already connected to the power grid. AI models' ability to create more accurate forecasts has started to impact this area. Sensors and smart meters collect real-time data across energy networks, and supervisory control and data acquisitions (SCADA) systems allow businesses to control processes locally or remotely, gather and monitor the smart meter data, interact with hardware such as sensors and motors, and record the resulting information. Data is also derived from forecasting weather patterns and atmospheric conditions and monitoring the amount of energy being produced at power plants and consumed.

There is also sufficient data related to customer consumption, customer behavior, and third-party data to allow AI algorithms to make successful usage predictions. This makes it possible to have a greater understanding of what is happening between power substations and usage meters. As a result, power companies today have a more granular view of the grid on a subhourly basis. Forecasts are becoming more accurate and stable, and self-monitoring devices that flag errors are providing operators with the information necessary to make better decisions.

An example of the potential of all this comes from Google Deep-Mind, which, in early 2017, revealed that it was in discussions with the UK's National Grid to help balance power supply and demand across Great Britain. The idea was to predict power demand surges by using data from weather forecasts and Internet searches combined with information about a customer's energy data to create a model that could be used to make the UK grid more efficient, saving energy, and avoiding overloads. Google expects that costs could be reduced by 10%.[11]

Energy companies have also been exploring demand-response solutions: the ability to have local customers lower electricity usage slightly at peak hours, so providers do not have to fire up what are known as "peakers" – that is, gas-fired plants that come online to fill in peak-demand situations and sell electricity at higher rates. When demand is even higher than what peakers can cover, energy companies often address this situation by buying and transporting power from other regions at higher costs. But a possible AI-based demand-response solution is already at hand in the form of the smart thermostat, Nest. Once Nest is installed in a home, the unit begins gathering data, learning when homeowners need more or less heating and cooling and adjusting systems to suit. This has not only saved Nest users money on heating bills; it has enabled utility companies to instruct Nest to, say, turn up thermostats slightly during a heatwave, when power is most expensive.

Austin Energy was one of the first to do this, offering Nest customers a small one-time rebate for the right to provide the service. This small demand-response solution has helped Austin avoid using ancillary generators, saving energy and money.[12] The changes made by Nest were so subtle that they generated no customer complaints, while reducing power demand. The same or similar solutions also enable energy companies to avoid spending the hundreds of millions of dollars it costs to build an additional power plant. This type of micro-optimization will become more important and perhaps even actively sought after in a more environmentally conscious world.

Oil Production

Energy companies, such as Shell and BP, are currently investing billions of dollars to enable new refineries, oilfields, and deep-water drilling platforms to utilize new AI algorithms.[13] AI models are being used in oil and gas exploration in a variety of areas, including seismic imaging and interpretation, well log analysis, production forecasting, and the analysis of reservoir properties.

Seismic imaging is a technique by which energy companies locate oil and gas reserves. Oil companies use equipment to produce seismic waves and then capture the propagation of these waves through the earth using sensors: essentially, using seismic imaging to take a sonogram of what is underground. Geophysicists study how these waves move through the ground and interpret this data to predict what is deep below the surface. After a human expert has labeled this data, AI systems apply supervised learning to detect similar conditions in three-dimensional seismic images. This both better targets drilling opportunities and allows for higher quality analysis, since AI models can consider all the data, whereas geoscientists can manually only use smaller portions of the data available to them.

When these oil or gas reserves are located, companies bore holes in the fields to determine the properties of those reservoirs, including critical information about subsurface rock formations. The bore-hole cores (1,000 – 1,500-foot section of the well) are extracted and analyzed in core labs to gather rock properties at millimeter scale resolution. Core lab analysis is based on various sensors, including CT scanning, X-ray fluorescence spectrometry, and others. Then, as wells are drilled, sensors are sent along with the drills into the ground, collecting data every few inches. This data is stored in well logs. Machine learning algorithms use this data from the core analysis along with well logs to predict rock properties for the new area of interest. Using this information, geophysicists determine the kinds and properties of rocks below the surface, including such characteristics as their density, porosity, and radioactivity. This information allows scientists to determine the presence and value of hydrocarbons in the rocks, and ultimately, the fraction of oil or gas that is recoverable from each field reservoir.

Data stored in well logs is also used to determine the direction in which to drill. Setting the drilling direction may result in simple vertical wells, but it also enables the successful exploitation of unconventional oil prospects. Unconventional in this case refers to certain types of hydrocarbons that are obtained in unconventional ways, such as fracking. These hydrocarbons need different extraction methods to free the oil. Drilling unconventional wells presents several challenges. To develop these reservoirs requires knowing optimal completion characteristics and stimulation methods for low-permeability reservoirs, as well as understanding the role of natural fractures in fluid flow.

The physics of fluid flow through various media can be extremely complicated and hard to model. To aid in getting the oil out of the ground, AI models are used for each of these steps. Data sets obtained from previously drilled wells, as well as data from third parties that include well header, logs, completions, and monthly production, is used to train AI models that predict the future performance of these existing wells using their history, as well as data from nearby wells that are drilled in similar geological locations. Fortunately, sensors are capturing more and more types of data at higher frequencies, including characteristics such as the density of the rocks, the electric resistance of the rocks known as resistivity, and the natural radioactivity of the rocks, which consists of gamma rays. This additional data has enabled AI models using deep learning to aid considerably in the ability of businesses to forecast oil prospects.

Even with well logs providing more data, there are often problems with the data collection itself due to malfunctioning sensors or adverse conditions within the borehole. In these cases, AI models fill in that missing data or remove bad data by identifying anomalies based on what algorithms have learned from previous data sets. Algorithms also use data sets from nearby or similar wells to provide geophysicists with a lot more data relatively quickly, helping them to determine where and how to drill. In order to determine where to drill, rock properties from core analysis and well logs are used to simulate hydraulic fractures, estimate the amount of hydrocarbons, and calculate the well trajectory. These models, for example, can help recommend that the drilling crew drill at a specified angle down, then turn around and drill horizontally to avoid harder rocks, and then drill down again in order to find the most productive locations. Enabled by AI, this knowledge of the properties of any given reservoir enables exploration and production teams to improve production forecasting, both in terms of determining a general production forecast and how oil pump parameters may affect that forecast.

BP took advantage of this ability in its attempt to exploit an aging Wyoming field. Partnering with San Francisco-based start-up

Kelvin, BP set up thousands of sensors across hundreds of wells. Kelvin's AI algorithms monitored the large streams of data coming from these wells, which allowed them to build a digital twin of the field. The twin was sophisticated enough to predict what the effects of opening a valve on one side of the field would be on the opposite side of the field, determining how pressure readings would change on the other side. When the system was operational, BP decided to let the AI system run independently, without human intervention. The company estimates that methane vented from the field is down 74% due to improved monitoring and maintenance enabled by the AI model. In addition, gas production rose 20% and costs fell 22%.[14]

Energy Trading

Companies in energy and commodities trading are continually improving their ability to rapidly detect market events and evaluate their effects on near-term market operations to discover new trading opportunities. They do this by developing a better qualitative and quantitative understanding of the impacts of these events. The goal is to realize incremental revenue and margin gain without needing to significantly grow the organizational or asset footprint.

Consider the simplified example of a global petroleum merchant, shown in Figure 5.3. Suppose the company has title for a vessel of gasoline in the US Gulf Coast and a commitment to deliver it at New York Harbor, ignoring complexities of some physical constraints and specification differences across regions. The company hears about an event: a refinery on the West Coast has shut down due to a fire. Company analysts believe that this will create a temporary price hike in California and would like to take advantage of this by diverting the vessel to the West Coast. However, before making this call, the analysts want to be sure that there will likely be a price hike in California and that the company will still be able to cover its obligation at New York Harbor without losing all the gains from the West Coast.

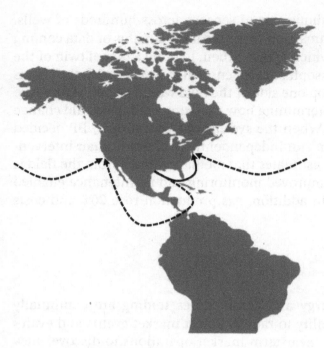

Figure 5.3   Energy trading scenario.

To rapidly evaluate these considerations, the company uses AI systems to see if there are other vessels destined for California, or if ships with a similar grade of gasoline could be diverted to California within a three-day window. If not, what is the range of the likely temporary price increase? The company also uses AI to discover if there are vessels that are available to acquire and direct toward New York Harbor within reasonable delivery ranges, pricing ranges, and demurrage, and what the total cost impact would be. (Demurrage refers to a charge that is payable to the owner of a chartered ship if there is a failure to load or discharge that ship within the agreed-upon time.) AI models provide the specific date and price ranges that would work to monetize the situation.

AI models are used to short-list the companies or groups to contact externally with the relevant requests, as well as allow the company to gain a better qualitative and quantitative understanding of the near-term markets and how it should respond.

Without a rapid and relevant analysis of the information, the company would not be able to capitalize on the event in California, because most events such as this are short-lived and require immediate action.

The assumption before now was that most events, like the refinery fire in this example, were highly visible. However, as companies began to compile their energy intelligence data, they realized that these kinds of situations, both large and small, were happening daily. AI systems have enabled businesses to identify new micro-opportunities, rapidly assess the impacts of more visible macro-opportunities as well as these newly visible micro-opportunities, and decide whether to take advantage of them.

While this chapter covers many examples of how AI is being used within the manufacturing and energy industries, this is an area in which AI usage is very nascent and far from saturated. In Chapter 12, we discuss the ambient sensing and physical control technical pattern that is relevant for these use cases.

In the next chapter, we will look at the ways AI's speed and flexibility has been aiding the healthcare industry.

Notes

1. *IEEE Industrial Electronics Magazine* 8, no. 2 (June 2014): 56–58. Industrie 4.0: Hit or Hype? https://www.researchgate.net/publication/263285662_Industrie_40_Hit_or_Hype_Industry_Forum (accessed September 26, 2019).

2. Obama White House (June 2011). Report to the President on Ensuring American Leadership in Advanced Manufacturing. Technical report, Executive Office of the President President's Council of Advisors on Science and Technology. https://obamawhitehouse.archives.gov/sites/default/files/microsites/ostp/pcast-advanced-manufacturing-june2011.pdf (accessed September 26, 2019).

3. Ibid.

4. *Forbes* (March 6, 2017). What Is Digital Twin Technology – And Why Is It So Important? https://www.forbes.com/sites/bernardmarr/2017/03/06/what-is-digital-twin-technology-and-why-is-it-so-important/#3ece60352e2a (accessed September 26, 2019).

5. Whirlpool Pro. Technology with a Purpose. https://www.whirlpool pro.com/connectivity/ (accessed September 26, 2019).

6. *Emerj* (August 13, 2019). Machine Learning in Manufacturing – Present and Future Use-Cases. https://emerj.com/ai-sector-overviews/machine-learning-in-manufacturing/) (accessed September 26, 2019).

7. US Department of Energy (August 2010). Operations & Maintenance Best Practices – A Guide to Achieving Operational Efficiency r3. Report No. PNNL-19634, section 5.4. https://www.pnnl.gov/main/publications/external/technical_reports/PNNL-19634.pdf (accessed September 26, 2019).

8. *Wall Street Journal.* How Manufacturers Achieve Top Quartile Performance. https://partners.wsj.com/emerson/unlocking-performance/how-manufacturers-can-achieve-top-quartile-performance/ (accessed September 26, 2019).

9. McKinsey & Company. Smartening up with Artificial Intelligence (AI) – What's in It for Germany and Its Industrial Sector? https://www.mckinsey.com/~/media/McKinsey/Industries/Semiconductors/Our%20Insights/Smartening%20up%20with%20artificial%20intelligence/Smartening-up-with-artificial-intelligence.ashx (accessed September 26, 2019).

10. H. Gleissner and C. J. Femerling (2013). *Logistics: Basics – Exercises – Case Studies.* Cham: Springer. (accessed September 26, 2019).

11. Ars Technica (March 14, 2017). DeepMind in Talks with National Grid to Reduce UK Energy Use by 10%. https://arstechnica.com/information-technology/2017/03/deepmind-national-grid-machine-learning/ (accessed September 26, 2019).

12. *MIT Technology Review* (May 20, 2014). The Lowly Thermostat, Now Minter of Megawatts. https://www.technologyreview.com/s/527366/the-lowly-thermostat-now-minter-of-megawatts/ (accessed September 26, 2019).

13. *Forbes* (January 14, 2019). How Algorithms Are Taking Over Big Oil. https://www.forbes.com/sites/christopherhelman/2019/01/14/how-algorithms-are-taking-over-big-oil/#269ed9428e2f (accessed September 26, 2019).

14. *Forbes* (May 8, 2018). BP's New Oilfield Roughneck Is An Algorithm https://www.forbes.com/sites/christopherhelman/2018/05/08/how-silicon-valley-is-helping-bp-bring-a-i-to-the-oil-patch/#bb1298430a89 (accessed September 26, 2019).

Chapter 6
AI in Healthcare

It's estimated that every patient will generate enough health data to fill nearly 300 million books in his or her lifetime. Meanwhile, research is expanding so rapidly that it would take physicians 150 hours a week to read everything published in their field . . . Machine learning has the potential to complement (not replace) healthcare providers and scientists.

Dr. Jonathan Lewin, CEO of Emory Healthcare, and Dr. Jeffrey Balser, CEO of Vanderbilt University Medical Center

The current state of AI in healthcare is a combination of brilliant successes, enormous potential, and a fair degree of frustration. The amount of data available to help physicians and medical researchers diagnose and treat illness is considerable, but existing systems can be fragmented and difficult to use. Computerized systems layered onto patient care were meant to free doctors from unnecessary work. Unfortunately, they too often add administrative burdens that reduce the effectiveness of patient-doctor interaction instead of increasing it. Some doctors report that they are spending as much as half their time trying to coordinate unconnected health solutions.

Despite that, however, there are already successful AI-driven systems in use meeting a variety of healthcare needs in areas including pharmaceutical drug discovery, diagnosis of illnesses, and hospital care.

Pharmaceutical Drug Discovery

Drug discovery is one of the more significant forms of research currently being conducted with the help of AI algorithms. Using data science and statistics in drug discovery is not new, but the chemistry used to make drug molecules is tricky. One reason is that molecules often contain a range of what are known as functional groups, consisting of one or more atoms that, when put together, produce the same or similar chemical reactions regardless of which molecule they are in.

To design a drug molecule, the important functional groups must be in the right places so the body can absorb the molecule, travel to the desired location (say, a tumor or infection site), have the desired interaction in that location, and then be eliminated from the body. But because drug molecules tend to be large and complicated and may typically contain several functional groups, it can be challenging to determine which of these groups to include as well as where to add them.

It is even harder to make a drug in the lab. Synthesizing anything chemical with this degree of complexity stretches science to its current limits. This is because to synthesize a drug, you combine smaller molecules, known as precursors, with other ones. Step by step, as you add more, these molecules grow larger and larger. Using currently available techniques, it is usually complicated to figure out a way to add a functional group in the place you want it without breaking what you already have.

It would save time, money, and effort to know theoretically if a drug can be created before trying to make it in the lab. That is why companies are currently using AI algorithms trained on massive volumes of chemical reaction data to suggest the best route to make these more complicated drug molecules. Although the use of AI has not shown many practical results yet, the area remains active and there are some early results that are promising. One example is how an AI model helped a pharmaceutical company radically shorten the drug development process, which typically takes seven to eight years, including three to four years of compound research. Using a *generative adversarial network* (GAN) AI model, the company was able to test 30,000 different compounds and determine a precise molecule that could be used for animal and human testing – shortening the research period from three to four years to just 46 days.[1]

Clinical Trials

Clinical trials cost quite a bit of money, and these costs are increasing. It can take up to $1 to $2 billion to make a drug and bring it to market. Trials are also time-consuming. This is particularly true of treatments for disorders whose progression is slow and hard to monitor, including neurological, neurodegenerative, psychiatric, degenerative, and aging-related ailments. The average initial drug discovery and preclinical testing takes three to six years; it then takes six to seven years for the three phases of clinical trials. After that, you need to submit the necessary data to regulators, such as the Food and Drug Administration (FDA) in the United States or the European Medicines Agency (EMA) in Europe, to be approved to sell the drugs. Then postapproval monitoring needs to be put in place.

Two of the factors that drive up the cost of trials are their duration and the number of patients needed to participate in them. Another is the fact that biomarkers measured during clinical trials to determine if a given drug is working may involve behavioral assessments by clinicians done by observation and the use of a stopwatch. For instance, a test might include counting how many times you can touch your nose and then the palm of your other hand alternately within one minute. Another might involve measuring how long a distance you can walk in six minutes. These somewhat primitive biomarker tests take a lot of time to perform and require large trial cohorts. Because they are subjectively measured, it makes them hard to evaluate without having large samples. Because many of the measurements are done manually by observing and using stopwatches, they are also imprecise – hence requiring a longer duration of the trial to measure the effect of the drug. These tests also require a patient to attend a clinic to be measured, rather than assessing her in her natural home environment. That means acquiring more patients and more clinicians, which drives up costs.

AI dramatically improves this situation by having trial subjects use wearable technologies like smartwatches or other sensors. These devices collect objective, quantifiable data that can take place during normal daily functioning. The result is better and more productive patient variables in the form of data that can be extracted and combined with other data on a particular disease in both smaller and more extensive clinical studies. Not only can these AI-enabled approaches reduce drug trial time and trial patient populations; they also reduce time to market, saving potentially hundreds of millions of dollars in the process. They do this by reducing the number of patients needed for the trials as well as precisely monitoring how treatments are working. In addition, the convenience for the patient reduces attrition, hence keeping the trials on time and budget.

Another costly aspect of the clinical trial process is finding and recruiting the patients for the study. It is a laborious process to find sufficient patients. Many trials get canceled because sufficient patients could not be recruited. To overcome this, more clinical sites at different locations are planned, increasing the cost of the trial. In Chapter 3, we discussed some of the techniques used to acquire a customer for e-commerce businesses by targeting relevant audiences for advertising specific products. Using the same methods, using data from social media, ad impressions, browsing behaviors, and other related information, patient recruitment success is being significantly improved for many trials. This is particularly relevant for trials of rare diseases where it is harder to find and maintain patients in trials.

Disease Diagnosis

For pathologists, images are an essential source of diagnostic and prognostic information. Here again, AI models are being put to use. Currently, there is a considerable amount of research being done into how machine learning techniques can be used to classify these images and provide quantitative information. One of these uses is in oncology image classification that identifies cancers.

Whole slide scanner technologies are already able to digitize pathology slides at microscopic resolution, making it possible to apply AI models to the images. Currently, researchers at Stanford University are training an algorithm to identify skin cancer, one of the most common types of cancer in humans. To do so, they labeled "nearly 130,000 skin-lesion images that represented more than 2,000 diseases to test whether the computer could distinguish harmless moles from malignant melanomas and carcinomas."[2] The results were noteworthy: their algorithm performed as well as a panel of 21 board-certified dermatologists.[3] The researchers plan to provide the tool to smartphone users in the future.

For many other disease cases, images have yet to be collected and labeled so they can be used as training data, because of the time it takes for doctors to label them as well as the roadblocks that are inevitable in such a highly regulated and fragmented industry. Even after labeling is done, it will take time for doctors to be able to use these AI solutions given that safety and efficacy must be proven to regulators before they can be implemented.

One critical area in healthcare today is determining the risk factors, diagnosis, and treatment of dementia. In a population in which living longer is becoming the norm, dementia is an increasingly serious problem: it is currently the leading cause of disability and dependence in older adults worldwide. A recent study done by scientists at McGill University used AI to develop an algorithm capable of recognizing the signatures of dementia a full two years before its onset using a single amyloid PET-scan of the brains of patients at risk of developing Alzheimer's disease. This algorithm should improve the way doctors handle patients as well as make clinical trials go more quickly and be more cost-effective, enabling treatments to come to market sooner.[4]

Another dementia study was done in 2018 by researchers at Boston University,[5] utilizing AI on data from the famous Framingham Heart Study. The researchers were using AI to discover the risk factors that increase the probability of getting dementia.

Unsurprisingly, age was a significant risk factor for the illness. However, the authors found other meaningful relationships in the data. Professor of neurobiology Rhoda Au wrote, "The analysis also identified a marital status of 'widowed,' lower BMI, and less sleep at midlife as risk factors of dementia." The hope is that the results will be useful to both individuals as well as clinicians.

Another use of AI and big data was Google's attempt to predict both heart disease and high blood pressure. The reason for testing those two markers is to be able to predict something even more severe: whether or not a patient will suffer a heart attack or stroke. To do this, the Google team examined a subject's retina, since medical researchers had previously noted "some correlation between retinal vessels and the risk of a major cardiovascular episode."[6] Using retinal images, Google says it was able to accurately predict which patient would experience a heart attack or other significant event within five years, and it was done accurately 70% of the time. These results are in line with tests that measure a patient's cholesterol levels in the blood.

Google's algorithm trained on data from 284,335 patients and tested on two independent datasets of 12,026 and 999 patients, respectively. Head researcher Dr. Lily Peng reported that the training dataset was admittedly small. However, she added, "We think that the accuracy of this prediction will go up a little bit more as we get more comprehensive data."[7]

AI is also currently being used in tracking and understanding the symptoms of Alzheimer's disease. Work on treating Alzheimer's patients based on an accumulation of data about their behavior is currently being done by researchers at MIT's Computer Science and Artificial Intelligence Laboratory (CSAIL).[8] To gather this data, the team has Velcroed a flat white box to the wall of a patient's room that can determine the biggest to the subtlest of patient movements. The machine tracks thousands of movements every day using low-power wireless signals to map things like gait speed, sleep patterns, location, and breathing patterns.

After this information is uploaded to the cloud, ML algorithms find patterns in the patients' daily movements, helping the researchers both diagnose and track the course of the disease. Tiny changes in the brain can manifest as subtle changes in behavior and sleep patterns years before Alzheimer symptoms show up. This enables the researchers to use AI models to identify patients at risk for developing the most severe form of the disease, which means earlier treatment, identifying patients who can benefit from experimental therapies, and even helping family members plan for eventual care needs.

Preparation for Palliative Care

Being able to determine when a patient will die is an incredibly valuable thing to know but notoriously tricky for doctors to decide. That is why AI is currently being used to predict the end of an individual's life. A study done in 2016 by a team at the Marie Curie Palliative Care Research Department at the University College, London, showed that doctor-predicted survival rates showed a wide error variation, anywhere from "an underestimate of 86 days to an overestimate of 93."[9] Why were the doctors so often wrong? Predicting mortality means taking into account a variety of complex factors, including family history, age, the nature of the illness itself, and the patient's reaction to a variety of drugs. Doctors' biases such as their natural reluctance to admit that the end is near, confuses the situation even further. Pulitzer-prize-winning physician and author Siddhartha Mukherjee has described the situation as follows: "Doctors have an abysmal track record of predicting which of our patients are going to die. Death is our ultimate black box."[10]

An accurate prediction of the time of death can make a tremendous difference to both patients and medical facilities. Palliative care can be administered at precisely the right time so that patients can take advantage of the different services available when they need them, including appropriate pain management, psychological support, and support for social, cultural, and spiritual needs. Good predictions also mean avoiding unnecessary burdens on the healthcare system itself, incurred when a patient enters care too early.

Could AI be used to determine precisely when a patient should be admitted to hospice? Connect with family and friends? Have drug regimens changed? That is what a research team from Stanford University was eager to find out. Using an AI algorithm to predict patient mortality, the team was hoping to improve the timing of end-of-life care for critically ill patients.[11]

The team reached its goals by using the hospital's medical records (data that existed before the patient's death) to figure out when it might have been best to be able to make an accurate prediction of when she might die. Doctors at the hospital had already coded this information, and it included everything from the patient's diagnosis to the number of scans done and the medical prescriptions written. Even though it was limited (there were no questionnaires or conversations, for example), it was fed into a deep neural network to see if the data could enable the algorithm to generate a probability score that a given patient would die within three to 12 months. The study used training data from approximately 175,000 patients and then was tested on 45,000 other patients. In addition to the hospital's medical records, the team added additional data from those patients that existed before that time to see if it would help predict the approximate date of death, as well as what kinds of inputs would teach an algorithm to make the prediction. The system proved accurate enough to predict the likelihood of death correctly in 90% of historical cases. This model is now used to provide doctors with a daily report of newly admitted hospital patients who have a high likelihood of death, for appropriate intervention or planning.

In another study, done by Yale University, a team of researchers using AI analyzed health data from a registry of more than 40,000 patients to try to improve the prediction of heart failure patient survival.[12] The study enabled the team not only to predict survival but also to enhance the choice of treatment options by grouping patients into four clusters, each of which had different responses to commonly used medications. This innovative approach could lead to better care for this incurable chronic condition.

Because of the black box problem, it is currently impossible to completely understand why AI is good at these predictions, so it is hard for doctors to learn from them. However, AI systems will likely continue to do a better job than physicians in some tasks. These types of AI models can help doctors to improve patient care, while reducing time-intensive stresses on medical professionals so they can bring their other impressive skillsets to bear.

Hospital Care

On January 24, 2018, Google published a research paper written by 34 of its AI researchers that claimed it had predicted outcomes of patient hospitalization with better accuracy than with existing software. Areas examined included whether a patient would be discharged and then readmitted; what her final diagnosis would be; and whether she would die in the hospital. In doing so, Google was able to accomplish something very significant: to do the study, it was able to obtain de-identified data of 216,221 adults, with more than 46 billion data points between them.

Google used three neural networks to learn from the data and work out which data was most useful to predict patient outcomes, identifying the words and events most closely identified with the three results. The AI algorithm used a previous Google project called Vizier and learned to ignore irrelevant data rather than relying on AI scientists to hand-select the data to be included in the analysis. Although the results have not been peer-evaluated, Google indicated that its neural nets were, among other things, able to predict patient deaths one to two days before currently available methods.[13]

AI can also help determine when patients can leave the hospital. A team at the Children's Hospital of Los Angeles (CHLA), is now using deep learning in the Pediatric Intensive Care Unit to predict acceptable states at discharge.[14] They do it with data gleaned from intensive care unit (ICU) encounters, each of which contains a wealth of information about disease progressions, treatments, and patient outcomes. "AI enables us to explore the information available from ICU encounters and generate models that understand complex relationships between drugs, interventions, and patient well-being," says David Ledbetter, a senior data scientist at CHLA. "In effect, we use AI to learn from those experiences and apply what's been learned to provide the best care possible for each child that comes through our doors."[15]

Machine learning algorithms require an enormous amount of data, and the groundwork with Madison, Wisconsin–based Epic, an electronic medical records company, provides exactly that. In 2018, Ochsner Health System, one of the first health systems in the country to integrate AI into patient care workflows, took advantage of this rich mine of data to launch an exciting new artificial intelligence tool. The ML platform, built by Epic and powered by Microsoft Azure, is meant to predict and thereby prevent "codes" – that is, those times when a patient goes into cardiac or respiratory arrest and needs immediate medical intervention. Ochsner ran a 90-day pilot study with the system late in 2017, and it "reduced the hospital's typical number of codes by 44 percent."

The platform tracks thousands of data points to predict which patients will deteriorate soon. Then it generates "precode" alerts that help Ochsner's care teams treat patients sooner and more proactively, saving time and lives. These alerts have been fine-tuned to provide a four-hour-ahead warning. "It is like a triage tool," says Michael Truxillo, medical director of the rapid response and resuscitation team at Ochsner Medical Center. "A physician may be supervising 16 to 20 patients on a unit and knowing who needs your attention the most is always a challenge. The tool says, 'Hey, based on lab values, vital signs and other data, look at this patient now.'"[16]

The Royal Free Hospital in the UK joined forces with Google's DeepMind to build an app called Streams to improve workforce effectiveness.[17] They focused on "failure to rescue" problems – that is, not getting the right physician to the right patient in time. Streams has so far been used to help clinicians better identify and treat acute kidney injury – a condition linked to 40,000 deaths in the UK every year, a quarter of which NHS England estimates are preventable. It is tricky to identify the symptoms, because it is a combination of small data signals that indicates this, and that those data points could mean many other things, and so are often missed. Streams brings together multiple data sources so they can detect them. Sarah Stanley, the nurse at Royal Free who leads the resuscitation team, said: "Streams is saving us a significant amount of time every day. The instant alerts about some of our most vulnerable patients mean we can get the right care to the right patients much more quickly."[18]

Another example of improving workforce effectiveness is the New York Presbyterian Hospital, one of the busiest hospitals in the United States. They have set up an offsite command center for nurses to monitor patients. The Clinical Operations Center (CLOC) monitors physiologic data coming from multiple sources in real time from smart-bed and other sensor-based technologies. The AI-assisted monitoring allows the CLOC nurses to direct how best the on site nurses should spend their time. This helps reduce alert fatigue for the nurses at the hospital and allows them to spend more time with patients. Leo Bodden, the hospital's chief technology officer said, "We have successfully been able to lean on highly complex automated systems that greatly decreased redundancy in tasks performed by registered nurses, doctors and other staff, reduced the number of team members physically required to monitor patients, and sizably cut down the amount of time staff spend inputting patient data."[19]

How are doctors and patients responding to AI's introduction to medicine? Admittedly, there is some resistance. Patients, for example, are worried that machines will replace their doctors, or that human beings will eventually be removed from all healthcare decisions. However, CHLA's David Ledbetter is sure that doctors will always be the final arbiters. So is Ochsner's medical director, Tuxillo. "Clinicians are inherently skeptical," Tuxillo's been quoted as saying. "They're wondering, 'Is artificial intelligence going to replace doctors?' The answer is no; doctors still must diagnose patients. The tool helps them prioritize care and synthesize the constant streams of changing information that doctors must track. This technology helps save lives. If we can make a difference for patients because of the alerts, we've done a tremendous service to them, their families, and the community."

In the same way that blood tests and MRI scanners did not replace doctors when they were introduced to the medical profession, neither will AI. It is merely another tool that doctors can use to probe, test, and analyze their patients, enabling them to provide superior care.

Notes

1. Technology Networks (September 5, 2019). Novel Drug Candidate Designed, Synthesized and Validated in 46 Days Using AI. https://www.technologynetworks.com/drug-discovery/news/novel-drug-candidate-designed-synthesized-and-validated-in-46-days-using-ai-323600 (accessed September 26, 2019).

2. Venture Beat (September 26, 2017). From Cancer Screening to Better Beer, Bots Are Building a Brighter Future. https://venturebeat.com/2017/09/26/from-cancer-screening-to-better-beer-bots-are-building-a-brighter-future/ (accessed September 26, 2019).

3. *Nature* (January 24, 2017). Dermatologist-level Classification of Skin Cancer with Deep Neural Networks. http://www.nature.com/doifinder/10.1038/nature21056 (accessed September 26, 2019).

4. McGill (August 22, 2017). Artificial Intelligence Predicts Dementia before Onset of Symptoms. https://www.mcgill.ca/newsroom/channels/news/artificial-intelligence-predicts-dementia-onset-symptoms-269722 (accessed September 26, 2019).

5. *Medical News Today* (May 10, 2018). New Dementia Risk Factors Uncovered. https://www.medicalnewstoday.com/articles/321747.php (accessed September 26, 2019).

6. *USA Today* (February 19, 2018). Google Hopes AI Can Predict Heart Disease by Looking at Retinas. https://www.usatoday.com/story/tech/2018/02/19/google-ai-can-predict-heart-disease-looking-pictures-retina/344547002/ (accessed September 26, 2019).

7. Ibid.

8. *MIT Technology Review* (March 19, 2018). AI Can Spot Signs of Alzheimer's Before Your Family Does. https://www.technologyreview.com/s/609236/ai-can-spot-signs-of-alzheimers-before-your-family-does/ (accessed September 26, 2019).

9. ResearchGate (August 2016). A Systematic Review of Predictions of Survival in Palliative Care: How Accurate Are Clinicians and Who Are the Experts? https://www.researchgate.net/publication/306922353_A_ Systematic_Review_of_Predictions_of_Survival_in_ Palliative_Care_How_Accurate_Are_Clinicians_and_Who_Are_ the_Experts (accessed September 26, 2019).

10. *New York Times* (January 3, 2018). This Cat Sensed Death. What if Computers Could, Too? https://www.nytimes.com/2018/01/03/ magazine/the-dying-algorithm.html?rref=collection%2F sectioncollection%2Fmagazine (accessed 26 September 2019).

11. *Stanford Medicine* (2018). Compassionate Intelligence. https:// stanmed.stanford.edu/2018summer/artificial-intelligence- puts-humanity-health-care.html (accessed September 26, 2019).

12. *Yale News* (April 12, 2018). Big Data Analysis Accurately Predicts Patient Survival from Heart Failure. https://news.yale.edu/2018/04/12/ big-data-analysis-accurately-predicts-patient-survival- heart-failure (accessed September 26, 2019).

13. Mass Device (January 29, 2018). Google Claims New AI-powered EHR Tech Can Predict Hospital Patient Outcomes, Including Death. https://www.massdevice.com/google-claims-new-ai-powered- ehr-tech-can-predict-hospital-patient-outcomes- including-death/ (accessed 26 September 2019).

14. *Journal of the American Medical Informatics Association* 25, no. 12 (December 2017). Predicting Individual Physiologically Acceptable States for Discharge from a Pediatric Intensive Care Unit. https:// www.researchgate.net/publication/321901871_Predicting_ Individual_Physiologically_Acceptable_States_for_ Discharge_from_a_Pediatric_Intensive_Care_Unit (accessed September 26, 2019).

15. Re-Work (April 9, 2018). How Can AI Help Improve Intensive Care? https://blog.re-work.co/how-can-ai-help-improve- intensive-care/ (accessed September 26, 2019).

16. Microsoft (April 9, 2018). Ochsner Health System: Preventing Cardiac Arrests with AI That Predicts Which Patients Will "Code." https:// news.microsoft.com/transform/ochsner-ai-prevents- cardiac-arrests-predicts-codes/ (accessed 26 September 2019).

17. Royal Free London, NHS (2017). Our Work with Google Health UK. https://www.royalfree.nhs.uk/patients-visitors/how- we-use-patient-information/our-work-with-deepmind/ (accessed September 26, 2019).

18. *The Evening Standard* (February 27, 2017). New Mother Receives Pioneering Kidney Treatment after App Detects Life-Threatening Illness. https://www.standard.co.uk/news/health/new-mother-receives-pioneering-kidney-treatment-after-app-detects-lifethreatening-illness-a3476936.html (accessed September 26, 2019).

19. Healthcare IT News (November 27, 2017). An Inside Look: NewYork-Presbyterian's AI Command Center. https://www.healthcareitnews.com/news/inside-look-newyork-presbyterians-ai-command-center (accessed September 26, 2019).

Part III

Building Your Enterprise AI Capability

Part III

Building Your
Enterprise
AI Capability

Chapter 7
Developing an AI Strategy

In the past, a lot of S&P 500 CEOs wished they had started thinking sooner than they did about their Internet strategy. I think five years from now there will be a number of S&P 500 CEOs that will wish they'd started thinking earlier about their AI strategy.
Andrew Ng, founder of Google Brain, founder of Coursera, AI professor at Stanford

As Andrew Ng notes, there were many business leaders who did not take seriously the fact that Internet technology was one of the primary opportunities in their businesses. Today many businesses leave the decision of how, or even whether, to use AI solutions to their individual business units. Often little action is taken beyond developing proofs of concept. When action is taken, major problems can arise when these units do not collaborate on AI strategy, leaving the company with a host of incompatible or competing AI implementations. Not only can this cost both time and money, but the resulting lack of standards can undermine the ability to transform at scale.

Fortunately, many companies and their boards are now making AI-enabled digital transformation a high priority. This is important to do as early as possible, since it is easy to underestimate how long change takes as well as how many challenges there are when implementing an AI strategy. Just because there is a trend in the market to use AI does not mean that the steps an enterprise should take are obvious.

It is natural for executives to worry about the difficulties inherent in integrating AI capability into a business and its culture, or about not having the right infrastructure to reap the benefits of AI algorithms. Even when integration and infrastructure are not in question, there is the issue of exactly which AI use cases and models are most useful and cost-effective for a business to implement. There is also the question of how to collect appropriate data. Having "big data" does not help AI implementations if that data is not relevant to the company's AI strategy. Without knowing which AI applications a business may utilize, data requirements remain unclear.

Despite all these difficulties, one thing is clear: organizations that properly scale their use of AI gain enormous advantages over competitors and businesses that do not struggle to compete. This chapter looks at how to develop an implementable AI strategy for the enterprise, with an eye toward addressing challenges that arise when no strategy is in place.

Goals of Connected Intelligence Systems

One way of thinking about how AI can benefit a business is to group its uses into three high-level categories: eradicating repetitive tasks performed by humans, generating insights and predictions, and amplifying human intelligence.

Eradicating repetitive tasks: Automating simple, repetitive tasks is sometimes referred to as "taking the robot out of the human," and computers have been good at it. As a result, the technologies of robotic process automation (RPA) can liberate employees from robotic work that requires little cognitive effort. Automation reduces costs, reduces human errors, and reduces employee dissatisfaction that comes from mindless labor in knowledge workers. RPA systems, which have been gaining traction since 2009, create orchestration layers: programs that link business processes already being performed by computers without having to rewrite all the software or require any substantial changes to existing applications. Unlike machine learning systems, these RPA systems are expert, that is, rules-based. That means RPA systems can be "taught" the keystrokes humans use to perform their tasks as well as the logic behind what the humans are doing. This setup allows businesses to replace humans who are performing boring, repetitive tasks (not jobs) with an automated RPA system. RPA alone is not always sufficient to complete a task, because there continue to be simple daily operational decisions to be made that the logical rules of RPA cannot currently handle. To enable RPA to make smarter or more complex decisions, it is combined with more intelligent AI systems, thus building the foundation for a digital workforce and intelligent robotic automation. This type of smart, decision-making solution has the advantage of not necessitating any changes to the underlying applications.

Generating insights and predictions: Machine learning and semantic reasoning systems give us the ability to automatically extract previously undiscovered insights from structured and unstructured data, enabling us to identify patterns in data and make predictions or find connections between and among events. From this, computers can recommend and perhaps even act on these recommendations. This leveraging of data improves a company's competitive advantage, enabling it to do such things as acquiring new customers, leading to new revenue, and reducing customer attrition and hence increasing margin and improving the supply chain.

Amplifying human intelligence: Systems that provide contextual recommendations for employees or customers leverage human judgment and creativity, working hand in hand with the analytical power of AI. We are already using applications to achieve this kind of augmented intelligence in localized ways, for example, where systems provide advice or knowledge in the context of a business process or user workflow (such as reading radiological images) and make knowledge-based tasks currently performed by employees more productive. The benefit of amplifying human intelligence is that it transitions businesses toward a form of business clairvoyance: when a situation occurs, managers and executives have already predicted that it might happen with some probability and are prepared to handle it appropriately.

The Challenges of Implementing AI

As indicated earlier, there are certainly challenges in attempting to implement AI initiatives. Business leaders often cannot easily decide precisely how and where to use AI in their organizations. Sometimes, the use cases they need to create value for their companies are not apparent. Too often, firms become stuck in the proof of concept (POC) stage. Sometimes, their studies show the potential value of implementing an AI use case, but management is concerned about putting AI into production because of worries about its "going rogue" and making bad decisions that could be harmful to the businesses from a regulatory, commercial, or brand perspective. At other times, there are collaborative challenges, in which the AI team is working in a silo and does not have buy-in to integrate its models into existing systems and business flows.

Other companies report that after commissioning projects, it took too long to get the data and the environment in place. Often, the data were in disparate systems, and managers did not have the budget or the skill (or both) to set up the high-performance computing environment that is often necessary for AI. Moreover, when it comes to profitability, companies that have managed to execute AI projects have reported mixed results at best. According to a 2016 survey, the average chance of making a profit from the implementation of a big data strategy was only 27%.[1] Enrique Dans has said, "It turns out there's a fatal flaw in most companies' approach to machine learning, the analytical tool of the future: 87% of projects do not get past the experimental phase and so never make it into production."[2] Some of the issues that get in the way of success are explained in more detail in the following subsections. Any successful AI strategy needs to address these challenges.

Hype versus reality: In AI, the gulf separating hype and reality is considerable. The term *artificial intelligence* has become somewhat of a catchall phrase, and without a deep understanding of the different types of artificial intelligence and how they work, it can be challenging to figure out what is real and what is not. This lack of clarity can make it hard to pursue a successful strategy. Firms may jump in without knowing what types of AI they should use and when they should employ it. Additionally, if the people making decisions about AI in an organization feel uncertain about its actual capability, value, or ROI, there may be limited or no adoption in their business.

Bad data: Even after they have started a project, many companies report problems with "dirty" data. These issues often show up as missing data, or incorrect values, or missing relationships among data entities. Not only do teams find out their data is not clean enough, they may discover that they do not have the data they need to move forward. There is also the challenge of duplicate data, which makes it challenging to have an accurate picture of, say, customers – costing time, money, and sometimes, reputation.[3] If supervised machine learning uses too much duplicate data, predictions may rely too heavily on those features most prevalent in the duplicates. This bad data, as well as a variety of other factors, can lead to more significant problems. For example, after a business has commissioned an AI project, it may take too long to get the data in place and cleansed and integrated with the AI modeling environment. This delay can result in the AI project taking too long to go live and show value, which leads to many projects being killed midway. Too many good ideas get abandoned at this stage.

Patterns of use: There may be many functional areas in a company where the use of artificial intelligence is desirable, but each area might have different needs. What HR wants out of artificial intelligence, which might be predictions about how successful new hires will be, can be very different from what the sales force finds valuable. On the other hand, there might be areas that appear to have very different needs but could actually be served by the same or similar algorithms. For example, predicting cardiac arrest in patients from historical heart-monitor data and predicting leaks in gas transportation pipelines are very different use cases, but they may utilize the same type of model and data structures underneath. A lack of understanding of this kind of pattern can cause a company to employ disparate and disconnected solutions. Without understanding the similarities among the different methods of AI and the potential for standardization and cross-leverage, managers may struggle to develop a comprehensive plan. The result is often duplicate efforts and higher cost.

Complexity and emerging technology: Mathematical modeling and data management are not accessible to everyone. Many of the models and algorithms in AI are inherently complex. Fine-tuning a model for the relevant decision variables takes skill, time, and patience. What can further exacerbate the situation is that AI technology is in constant flux, taking on new challenges and breaking new ground. There are not only a variety of different approaches and tools currently available, more are continually appearing. This proliferation can make it challenging to know which to use and what has become obsolete. Between AI's complexity and the constant introduction of new technology, it can be difficult for firms to proactively pursue a sustainable AI strategy, forcing many to implement within organizational silos or defer AI solutions altogether.

Evolving regulation: Both existing and upcoming regulations are beginning to change the digital marketing landscape. In 2018, General Data Protection Regulation (GDPR) went into effect. This legislation was the result of almost four years of work that began in January 2012, when the idea of data protection reform across the European Union was introduced. In 2020, the California Consumer Privacy Act (CCPA) became law. Both sets of regulations aim to ensure that data protection and privacy be upheld for their respective citizens. In the case of the GDPR, it protects citizens of the European Union (EU) and the European Economic Area (EEA) and affects companies doing business in those areas or with citizens of those areas. In the case of the CCPA, it protects California residents. Protected individuals are entitled to know what data is collected about them and if and when that data is sold or transferred. They can also access their data, require that businesses delete personal information about them, and prohibit the sale of their personal data. These regulations, as well as future ones, will affect not only those protected but businesses around the world, and companies are already striving for greater transparency and tighter controls over personal data. We discuss emerging regulations and policies further in Chapter 14.

Scarce talent: AI talent is currently in short supply. Although estimates vary, in December 2017, Chinese Internet giant Tencent Holdings Ltd. guessed that the world has open positions for perhaps 200,000 to 300,000 AI practitioners and researchers.[4] Anecdotally, this statistic is reflected in both the high salaries commanded by AI scientists and long recruiting times. Given the numerous options that potential candidates have, they can be choosy, and they often prefer to work where they can be part of a strong team and have a variety of problems to work on; where clean data is readily available; and where a robust platform for modeling already exists. Without these factors in place, companies find that recruiting top-tier talent is very difficult.

Misaligned execution: Even when a company has managed to hire the right AI team members, a team may build an AI model that solves problems that the business does not consider a priority, does not know how to use, or does not trust. This can happen for a variety of reasons. The IT department – a critical part of the scaling of any AI solution – may not have supported its development and deployment or may have allotted insufficient funds to it. The infrastructure team may not have had enough of a business case to requisition the use of the necessary servers or infrastructure to support it. Data governance may have restricted the use of necessary data due to confusing regulations, or data engineering may have delayed building adequate data pipelines due to cumbersome processes. Or it can happen because the business users were not involved enough in the model definition and development process. Even more fundamentally, the AI science team either may not understand the business well enough to create appropriate algorithms or cannot articulate a solution in a way that gets users on board. Management itself may also have neglected to articulate specific needs and goals, leaving things to the AI team, or refused to exercise enough influence to compel users and IT to support the team. A business may also simply fail to implement good solutions, pushing them off into an indefinite future, if these solutions change business processes too much or require additional investment.

AI Strategy Components

A root cause of many AI implementation problems is not having developed a comprehensive, cohesive, and sustainable AI strategy and roadmap. As executives begin to think about leveraging AI in their companies, they must decide what they want to get out of it. Thinking through the following deliverables can help shape such a strategy.

First, the business should set *goals* that define the purpose and vision of AI within the firm. The goals may include sustained competitive advantage, incremental revenue opportunities, or cost reduction. Often the goal is a holistic AI transformation that can support digital business transformation and business innovation through new products, services, or business models.

Next, the business should identify *use cases* that define the potential short-term and long-term ways it can use AI to drive its goals. Use cases help determine how the company's data is monetized. This use case catalog allows the business to accomplish a variety of objectives, enabling it to understand the overall business case for AI and showing the different ways it can provide value. Identifying use cases helps define holistic requirements for the types of AI solutions desired as well as the associated information management to support these solutions. It also helps the business understand the user groups and business processes that may be affected. This information is needed to create the roadmap and change management plans.

The business should then create an *architecture* that encompasses the technology components and platforms that the company will use to support its AI and data needs. This architecture involves not only the AI components but how to make the enterprise architecture and existing applications AI-centric. It should include information gathering, storage and processing, AI modeling, visualization, user experience, model management and deployment, and integration with line-of-business systems or business processes for activation.

Firms also need a *data strategy* and data readiness plan in place. There should be ways to make relevant data for the use cases available, and that data should be of appropriate quality and timeliness. Data readiness may include activities such as data quality assessment and remediation and data lifecycle governance.

Businesses must establish the *organizational capability*, including the structure, talent, and processes required to execute AI projects at scale. Organizational capability includes decisions around the structure; for example, if AI should reside within one group, be embedded in relevant business groups, or maintain a community of practice. It should also include the roles and processes to sustain the AI lifecycle, the necessary skill sets needed to nurture and grow it, and how to maintain a connection with the external AI ecosystem.

Lastly, it must establish a system of *governance and change management* that defines how decisions are made, how AI model "safety" is maintained, and how the company deploys and utilizes this new capability enterprise-wide. The governance structure determines such things as project owners, standards, value measurement, and project approval and prioritization. Governance also defines the research agenda in terms of market analysis, competitive assessment, and vendor assessment. Change management works on bringing the leadership and the company along on the AI journey to increase active engagement and reduce resistance within the organization.

Steps to Develop an AI Strategy

Once the preceding deliverables are determined, it is time to take the essential steps toward extending and amplifying the vision that will make the organization's AI plans a reality. First, the executive team members must be brought up to speed, so they collectively have a similar level of understanding and appreciation for AI. They need to understand what is essential and achievable and establish consensus on the focus areas and objectives of an AI program. Having unrealistic expectations can sink the best-intentioned AI effort.

Executive briefings should include investigating the credibility of AI: a critical look at the hype versus reality, informed by how AI has benefited enterprises across a range of industries. The briefings should give everyone on the team a clear vision of what successful use of AI looks like.

After the briefing is an excellent time to think about forming an AI team, because each next step takes more effort than the previous one. Team formation may be carried out by initially bringing together like-minded people from inside the organization, and then augmenting this group with external collaborators and new hires. Decisions to be made by this team include ownership of activities such as driving use case discovery and defining the architecture needs of the AI system.

Next, the business should identify the specific areas of opportunity. Alignment needs to be reached on the critical drivers for engaging (or improving) AI technology in the company. Where will it work best? Is it worth the investment of time and money? Identifying opportunities should include examining potential use cases for the company and understanding how AI in those areas can enhance the business as a whole. For each area, determine business objectives, anticipated returns, impacts on current processes and systems, and the achievability of goals. Prioritize those areas of opportunity. Determine which of the aforementioned business cases are the most compelling, impactful, and viable for the enterprise.

After this, the business should determine its readiness to implement AI in the areas of opportunity identified earlier. The assessment should include data, technology, organizational structure, governance, brand, and supply chain and partners. Business leaders need to understand how prepared they are to exploit AI technologies across a range of technical and nontechnical dimensions. They must figure out the gaps between the current state of affairs and the required capabilities for an AI system based on both this assessment and an understanding of comparable enterprises making similar transitions, even if they are from other industries. Understanding the gaps helps the business identify areas for improvement and define a transformation path.

Once all these steps have been taken, the business can develop a strategic and operational AI roadmap to help move the organization toward using AI at scale. The company can determine the activities required and the approach, priorities, high-level plan, milestones, and timetable based on the risk profile, sense of urgency, level of preparedness, and budget and schedule considerations. This roadmap should cover any gaps in the readiness assessment discussed earlier and ensure that each deficiency includes plans for remediation.

Plans at this level do not allow for a detailed assessment of financial business impacts, but businesses can begin to map out what they need to make a high-level business case. The business case includes a qualitative or order-of-magnitude assessment of costs and benefits. Potential costs to the organization may consist of the need to prioritize or reallocate resources, as well as making required investments in staff, training, software, hardware, cloud/hosting, consulting, and operations. Where possible, trade-offs and associated benefits and costs of the opportunity should be identified. Potential benefits to the enterprise of using artificial intelligence may include improvements in operational efficiency, cost reductions, revenue generation, improved time to market, IP value, competitive advantage, customer satisfaction, and increased market share.

At this point, it is appropriate to start getting people within the organization comfortable with AI. Utilizing the same tools as previously employed with the executive team, the broader organization can be taught to understand what is essential and achievable. Leaders should plan educational roadshows throughout the company to get everyone up to speed and move the company toward an AI and data-driven culture.

Some Assembly Required

The remainder of this part of the book covers in more detail some of the topics that are needed to define an AI strategy and transformation roadmap. There are a few fundamental elements that are important to ensure that the new AI enterprise can scale, delivering transformative value while being worth the time, effort, and money it takes to get there. These include a transformative data strategy, an AI platform, and an AI center of excellence to sustain the program.

Creating an AI Center of Excellence

The most effective way to scale AI across a company is to set up an AI center of excellence (AI CoE). An AI CoE is the organization in a company that executes an AI plan. It helps define and maintain the strategy for the platform, data, and talent across business units. It is also the place where decisions about governance and management of the AI platform are made.

There are three different execution models for where the AI teams sit (see Figure 11.1 in Chapter 11). Most companies will likely decide on and move toward a federated model. In a federated model, the AI talent and scientists do not all sit within the CoE. Instead, those employees work within individual business units (BUs), with the CoE supporting them as needed. This setup enables specific BUs to innovate locally where appropriate, generating their particular use cases while leveraging the CoE to collaborate across BUs when appropriate. If a model is overly centralized (shown at the right of Figure 11.1), then local use cases do not get the attention they deserve. If a model is too decentralized (shown at the left of Figure 11.1), then there might be too little cooperation among BUs, forcing each unit to set its own strategy, resulting in redundancy and inefficiency. The upshot is that everything going into actual production comes at a relatively higher cost, providing lower returns and taking longer to achieve the business case.

Building and maintaining relationships with peers, universities, research institutions, and product vendors is essential in the AI ecosystem. These relationships enable the CoE to keep up with what others in the industry are doing and lay the path for leveraging academic research and recruiting. These relationships are particularly critical because the field is moving so quickly. Appropriate technology partners that enhance the ecosystem are also essential.

These partners can have a direct impact on the pace and scale of adoption of AI solutions within the organization. Additionally, the right technology partners can help the business understand where the industry is likely to be headed. The AI landscape is evolving rapidly; it is necessary to remain abreast of the state of the industry to choose the best suppliers, vendors, and collaborators for an enterprise. The AI CoE functions and their interactions with other parts of the company are covered more fully in Chapter 11.

Building an AI Platform

An AI platform is the hardware, software, and tools framework that enables a business to accelerate the full lifecycle of enterprise AI projects at scale. When companies create AI projects without robust platforms, they waste time and money before they get any significant returns. The worst part of this is not that a project takes longer and has a lower return on investment; it is that people lose heart and abandon ship. The numbers are not working, so the project is killed, inevitably leading to wasted money and lower credibility for future projects. Having a platform and its associated processes allows for faster cycles from experiment to production and automates much of the grunt work for the users of the platform.

There are a few key things that an AI platform should provide. It is essential to base the AI system's design on requirements from across the entire business, both as it is now and anticipated to be into the future. The design should support the lifecycles of AI and machine learning projects, from data management to model experimentation to moving working models into production and monitoring them on an ongoing basis. Additionally, the design should create self-service for the AI scientists without being dependent on IT: the data and the processes should be accessible for them to use. Moreover, it should cover the breadth of AI and machine learning tools, frameworks, and methods that AI scientists are likely to use so that waiting on infrastructure or procurement does not hinder their progress. The AI platform is covered more fully in Chapters 9 and 12.

Defining a Data Strategy

In the past, data was considered simply as the byproduct of a business procedure. Although such functions as customer service and financial reporting might need to access the data for follow-up, it otherwise had little value and was often not easily accessible after its initial use. Within the past 20 years, however, things have changed dramatically. Until recently, subject matter experts embedded their knowledge in the logic they built, creating the software for most decision-support applications. With machine learning, computers derive logic from data without explicit human intervention. This means data has become of paramount importance within the organization.

Today, everyone agrees that data is a highly valuable asset for an organization, fundamental not just to decision-making, but also for creating or refining products and services. In fact, most leading businesses today are moving toward data-driven and algorithmic strategies. However, although data may be a vital source of insights, a handful of large companies control vast amounts of it, particularly personal data. This near monopoly is not the case with algorithms; no one owns the mathematics of AI. Open-source frameworks are better than ever, and many of them are supporting a variety of AI initiatives. Although high-quality machine learning packages are also proliferating, high-quality data remains scarce. However, despite this, few companies have actually adjusted the way they capture, store, and manage their data estate. A data strategy – a set of rules, policies, and standards regarding the management of data across a company – is essential to ensure that maximum value can be extracted from the data.

A modern data strategy allows companies to develop a balance between tighter *control* and greater *flexibility*. With increasing regulatory and reputational concern, many companies are putting together data strategies around standardization and compliance. Control and standardization of data have historically been required for compliance and regulatory reporting and continue to be expected. But now, flexibility and data sharing are much more critical to driving business decisions and the development of new products and services. Although most companies understand the approach for data control, standards, and quality through data governance, many are just starting to think about the flexibility, sharing, reuse, and monetization necessary to transform data into a strategic asset.

As part of their data strategy, companies need to decide both their *data monetization* strategy and their *data commercialization* strategy. Data monetization uses data within the company to generate value, including consumption of third-party data. Monetization is through implementing algorithms for the use cases mentioned earlier, using the data internally within the company. Data commercialization is creating products and services for other companies based on data generated or aggregated within the company. This becomes a new revenue stream. Increasingly, more companies are leveraging data commercialization as part of their data strategy.

Data must be high quality and as comprehensive as possible to create a data-driven competitive advantage. Businesses should look at the data from the perspective of creating a digital twin of the business – of the customers, products, processes, and the business environment. If the data comes from the company itself (first-party data), it has to be aggregated and cleansed. Determinations must also be made about what additional data is needed for modeling and where to get it. Sometimes, a business can gather this additional data without outside intervention. For example, a bank that is employing customers' sociodemographic profiles to make offer recommendations can increase responses significantly by using behavioral data it already has but is not using in the offer process. This data is collected when customers interact with its website, apps, content, and media.

To paraphrase a quote attributed to Bill Joy, co-founder of Sun Microsystems, no matter how much useful data you have in the company, there is more useful data outside your company. This is in the form of second- and third-party data. Second-party data is what a business agrees to acquire from their partners, suppliers, and vendors. A company might, for example, ask partners for more granular sales data to enable better forecasting of their demand. Third-party data is data that can be purchased from independent sources. This data includes such things as geolocation data or satellite imagery to enable an understanding of, say, foot traffic or parking lot usage at retail stores. Data obtained from each of these sources enhances a company's ability to create models in new areas and with better accuracy. In these ways and others, data and machine learning can be used to increase an understanding of the business, improving and maintaining a competitive advantage. Figure 7.1 shows the various types of second-party data that many companies are aggregating. They are doing some of this directly by scraping websites. In addition, hundreds of companies have emerged that are generating or aggregating this kind of information at a level of quality that is easily purchased and more seamlessly integrated.

To harness all this data successfully, governance needs to be established. Data governance ensures that high-quality data exists throughout the enterprise for when it is required. Is it usable? Is it accurate? Who is responsible for it and who is allowed to see it? Is it secure? Data governance, security, and privacy are relevant topics that have been extensively discussed in other books, so we do not cover them here in depth.

Moving Ahead

Once goals are established, executive buy-in is obtained, talent as well as data are being gathered from throughout the organization and externally, employees are engaged, and an AI CoE is under construction, it is time to look more closely at how things will work. In the next chapter we discuss the workflow for how to implement an AI project in an organization. Understanding this workflow will help explain what systems, risk management policies, and organizational structures can best support the AI implementation lifecycle.

Firmographic	Demographic	Technographic	Relationship	Satellite	News and Blogs	Device	Behavioral
Revenues	Biography	IT infrastructure	Social network	Foot traffic	News sites	Reverse IP address	Content consumption
Employees	Title/role	Applications (used or installed)	Reporting relationship	Parking lot usage	Analyst and broker reports	Geolocation	Search history
SIC/NAICS codes	Time in title/role	Professional services contracts	External social contacts	Inventory and movement	Blogs	Usage	Site, page, and app visits
Office locations	Office Location	Equipment	Internal colleague and employee contacts				Social posts
Department and business unit	Department	Other technologies	Influence rating (Klout or proprietary score)				
Press releases	Education	Contract renewal date					
Job postings							

Figure 7.1 Different types of third-party data that is available commercially or through the web.

135

Notes

1. CapGemini (2016). The Big Data Payoff: Turning Big Data into Business Value. https://www.capgemini.com/wp-content/uploads/2017/07/the_big_data_payoff-turning_big_data_into_business_value.pdf (accessed September 27, 2019).

2. *Forbes* (July 21, 2019). Stop Experimenting with Machine Learning and Start Actually Using It. https://www.forbes.com/sites/enriquedans/2019/07/21/stop-experimenting-with-machine-learning-and-start-actually-using-it/#c6c625933651 (accessed September 27, 2019).

3. QGate. 10 Reasons Why Duplicate Data Is Harming Your Business. https://www.qgate.co.uk/blog/10-reasons-why-duplicate-data-is-harming-your-business/ (accessed September 27, 2019).

4. The Verge (December 5, 2017). Tencent Says There Are Only 300,000 AI Engineers Worldwide, but Millions Are Needed. https://www.theverge.com/2017/12/5/16737224/global-ai-talent-shortfall-tencent-report (accessed September 27, 2019).

Chapter 8
The AI Lifecycle

Our success at Amazon is a function of how many experiments we do per year, per month, per week, per day . . .
Jeff Bezos, CEO of Amazon

This chapter provides a high-level AI model creation workflow: a description of the steps needed to execute an AI project at scale, from finding use cases to getting the model deployed and used in production. Some of the steps are iterative and may overlap, but in general, they take place in the order described. Understanding the difference between a variety of machine learning solutions can be challenging for those who are not AI experts, and there may be many algorithms and a variety of ways that can be used to solve business needs. This does not mean you have to become an expert in every modeling nuance to manage an AI team or project successfully. However, it is essential to understand not just what the steps in creating each model are but why they are necessary, allowing team managers to distinguish between what is working and what is not. Additionally, understanding this lifecycle makes clear why the AI platform features described in the next chapter are necessary.

Figure 8.1 shows the high-level end-to-end process for modeling. The first step in the process is to identify and define use cases. For each use case, the team decides what specific questions they

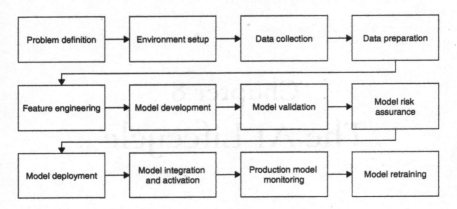

Figure 8.1 The workflow for AI, machine learning, and data science projects.

want answered and determines how answering these questions benefits the business. It is crucial to resolve decisions early on about how to frame a use case and define its qualitative benefits, target metrics, and data needs. The next step is collecting and assessing the data. What data is available? How much is there? How good is it? Where is it currently housed? How easy is it to access? Is there signal in the data? Is there a need to acquire additional data from inside or outside sources, and how easy is it to get? The team then remediates any issues with the data, cleanses it, and gets it ready to create a baseline learning model.

The team creates the best model or models for its needs, training an AI algorithm that is most appropriate for the use case, the data available, and the questions that the business wants to answer. The result might be one model or a number of them that can be used together. Finally, the team tests and prepares for deploying the model for use in production. Let's look at these steps in greater detail.

Defining Use Cases

Identifying use cases generally starts with understanding the actors within an organization. The actors are often employees, but they can also be customers, partners, or suppliers. For example, in a retail company, the employees are people working at the store or in finance; the customers are people coming to the store or online to purchase; the suppliers are the people making or delivering the products to the store to be sold; and partners may be people that pick up products at the store for delivery (such as Instacart). These individuals are all considered actors. Creating an actor map – a high-level document describing who the actors are and what their primary functions or goals are – is a good starting point.

Next, the team looks at the business processes that each team of actors utilizes to achieve their function or goals. The nature of each task is examined and the activities grouped into categories, such as what the actors do daily or regularly versus things they do sporadically; jobs that do not require complex decisions versus those that require patient deliberation; and actions that are part of a collaborative process versus things each actor does independently. Would any additional information or knowledge improve the function or goal, or enhance part of the process or possibly eliminate it altogether? Does it include predicting something? Would predicting something help an actor perform a task? What would need to be anticipated? Are the existing prediction goals well defined and quantified? For example, if a telecom company could predict which customers are most likely to cancel their phone service, they would be able to focus their preventive actions on just that set of customers.

An excellent way to explore business practices is by asking key questions of the actors involved. These might include questions such as:

- What do you wish you did not have to do because of its repetitive and uninteresting nature?
- Something has happened. What do you wish you knew about it that you are finding difficult to research?
- What do you wish you could predict – in other words, what is likely to happen even though it has not happened yet?

Responses to these questions and others help define what type of AI task is likely to address the actors' needs. Examples include information extraction from unstructured text using natural language processing (NLP), predictions using machine learning or deep learning, action automation using RPA, or some combination of the three.

Having gone through this process, the team has determined a broad list of potential use cases (see Figure 8.2 for a sample "decision space"): what the use-case objectives are, what the modeling tasks are, what data is required, what business process or function does it fit within, what actors does it impact, and so on. The next step is to prioritize these. Additional information is needed for each potential use case so that it can be listed and acted upon appropriately.

It is worth noting that when a team is setting up to start on a project, it is often not clear if a particular use case can be implemented. AI is called a science because the team creates hypotheses and then tries to validate or invalidate these hypotheses. The mantra often heard on the conference circuit of "fail fast" is really about how quickly and with how little effort a team can determine if a "hypothesis" will not work. The more experiments the team can execute, the faster it finds useful models, and the higher the value it realizes. This is what Jeff Bezos is referring to in the quote at the beginning of this chapter.

The use cases selected should first and foremost be relevant to the success of the organization. To determine this, we generally look at two attributes – complexity and value. This can be done in two passes: first, a coarse-grained pass to understand relative prioritization, and then, for the selected set of use cases, a more fine-grained pass at the business value and cost, including the business case and ROI. We cover the coarse-grained aspects later.

It is essential to get a sense of the relative value of each use case. To understand its value, the team looks at how the AI task (e.g. automation, knowledge extraction, or prediction) could improve business outcomes. For example, if AI is used to make a prediction, but the prediction is not actionable – if there is no way to use that information to affect a business process – then there is no value that can be generated from it. If the AI task is actionable, then the team looks at impact: for example, if it is automation, are we looking at

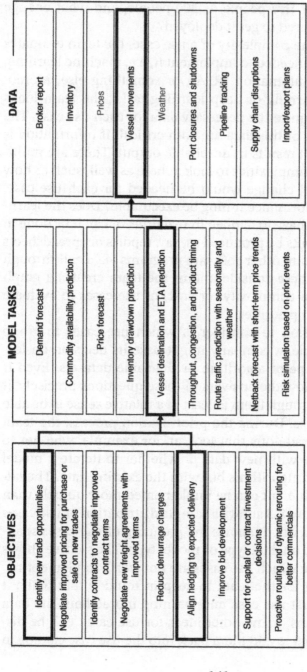

Figure 8.2 A sample map of use case objectives, modeling tasks to support the decisions, and data required to support modeling.

The figure contains three columns:

OBJECTIVES
- Identify new trade opportunities
- Negotiate improved pricing for purchase or sale on new trades
- Identify contracts to negotiate improved contract terms
- Negotiate new freight agreements with improved terms
- Reduce demurrage charges
- Align hedging to expected delivery
- Improve bid development
- Support for capital or contract investment decisions
- Proactive routing and dynamic rerouting for better commercials

MODEL TASKS
- Demand forecast
- Commodity availability prediction
- Price forecast
- Inventory drawdown prediction
- Vessel destination and ETA prediction
- Throughput, congestion, and product timing
- Route traffic prediction with seasonality and weather
- Netback forecast with short-term price trends
- Risk simulation based on prior events

DATA
- Broker report
- Inventory
- Prices
- Vessel movements
- Weather
- Port closures and shutdowns
- Pipeline tracking
- Supply chain disruptions
- Import/export plans

automating a job for five people or 300 people, and what change management is required to get it deployed?

To understand the complexity of a use case, the team evaluates the type of AI that is needed to implement it, e.g. machine learning, robotic processing automation (RPA), or something else. Moreover, for machine learning or deep learning tasks, the team looks at whether it can acquire the necessary data for each use case. The quantity and quality of this data are both critical. If information is not accurate, using it results in unreliable output. There are multiple other potential complexities to look at here as well, such as how much organizational change would be needed for each use case. Will testing and performance tuning be executable? Does the learning or the activation have to happen in real time, or can it be done in batch? Will the outputs be produced as, say, reports or spreadsheets for users or provided to other software programs via a call through an API? Based on these considerations, the team creates a graph of relative value and complexity (or cost) of use cases to evaluate which are the higher priorities.

Another dimension to consider is which use cases to group together. For example, a company might base its decision on leveraging the same type of algorithm or the same datasets. Even if managers do not know the answers to these questions precisely, it is crucial to think through them at least in a relative sense to be able to prioritize use cases. During the prioritization process, the team is often managing trade-offs that look at, for example, whether to start a new use case with new data or whether to iterate a model with more advanced algorithms but with the existing data. The latter may be lower value but can be implemented more quickly than the new model. Figure 8.3 shows a sample distribution of use cases by value and complexity. It shows the grouping of use cases spanning priorities based on data type or model type. The s-curve shows the ROI cutoff above which use cases are not worth implementing.

One more thing for the team to keep in mind is that while it is developing the first use case and learning more about the data through the iterations, many other ideas for use cases will be discovered. It is essential to add them back into the use case catalog in

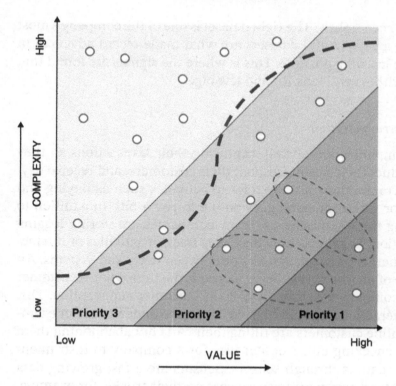

Figure 8.3 Graph showing use cases by value and complexity.

the backlog. These use cases help create incremental value; they are also part of the reason why building an a priori business case can sometimes be challenging. Teams often discover about four new use cases for each one they implement.

Collecting, Assessing, and Remediating Data

Collecting, assessing, and remediating data is one of the more critical steps to take toward a successful AI use case implementation. It is essential to accumulate as much relevant data as possible about the organization's customers, products, transactions, and processes. There is often a tremendous volume of data that has already been collected, but is there enough data relevant to what the business wants to accomplish? The right dataset is one of the company's most valuable assets. Useful datasets are what made recent advances in machine learning possible. This is where the signals are found that the algorithms will look for and leverage.

Data Instrumentation

Most companies are already capturing their transactions so they can conduct their businesses, bill their customers, and receive payments. Transactions usually refer to activities such as buying and selling, or when someone goes online to pay a bill. In addition to capturing these transactions, many companies are storing logging information. Logging is how they keep track of activities or interactions, whether between systems or between users and systems. An example of an interaction being logged is the time when a customer visits a company's website or calls its customer support line. This user-generated data might help businesses to understand, for example, if online customers are filling their carts but abandoning them and not checking out. Another way for a company to instrument for more data is through sensors. Sensors are a fast-growing data source. Many companies use sensors on their trucks, for example, to measure the positions of these trucks in their fleet.

For each use case under consideration, the team looks at data already collected and how it has been collected. If the application is user-facing, team members figure out if the system is logging all relevant user interactions. It is also a good idea to determine how easy and worthwhile it would be to record additional interactions that might be useful but are not yet instrumented. *Data instrumentation* refers to capturing and measuring data about processes as they occur. The more instrumentation that is done, the more data there is. In the earlier online shopping example, additional instrumentation might be to capture keystroke dynamics and mouse and touchpad movements (which would have to be instrumented on the client-side or in the app).

Using transactional data, AI models make recommendations to the user. By using incremental data about a user's clickstream, what she is browsing, what she added to the shopping cart, or what she removed, AI models "understand" why a customer may have abandoned her shopping cart – for example, to explore other products or because she lost interest. This helps refine the appropriate message to bring her back rather than trying a generic retargeting campaign. With additional data about a customer's keystrokes and mouse movements, AI models get an early understanding of whether and why a given customer is likely to abandon her cart so that the appropriate intervention can be tried. With each layer of additional data that is instrumented and used, different types of predictions and responses become possible, and the overall conversion continues to improve at each step.

As discussed in previous chapters, not all data needs to be collected internally. Third-party data coming from outside an organization can be purchased. One often-used example of third-party data is digital ad-impressions, which are used in customer targeting and personalization. Other examples include satellite images of parking lots to understand customer trends, geolocation information, and competitive trends from the news. Over the last several years, a large number of companies have emerged that offer many types of alternate data. When evaluating AI use cases, teams not only look at existing datasets, but also assess what other datasets should be instrumented or acquired from third parties to best serve their needs.

Data Cleansing

Data in its raw form is seldom ready for use in modeling. It needs to be carefully prepared to make sure nothing is wrong or missing. If something is, someone needs to remediate the data through *data cleansing*, one of the most important and underappreciated parts of data-driven modeling. Data cleansing is essential to getting an AI model off to the right start. It is the way to discover and fix data issues such as bad data collected due to unreliable sensors, data that is unavailable due to a software change that has broken a logging mechanism, or misinterpreted data. Missing data can result in weak predictions, and deciding how to deal with missing data is part of the cleansing process. Different models may be more or less sensitive to missing data, and even models that ordinarily handle missing data well can sometimes be sensitive to it. When it is necessary to fill in data that is missing, teams can use models for missing value imputation.

Organizing the data is also essential. A company may have aggregated a lot of data, but an easy-to-use and reliable *data pipeline* is necessary when putting an AI platform into place. The flow includes where to store the data, how to get it there, and how easy it is to access and analyze once it is there. The team must also check the robustness of the data stream. Are external sources accurate? Are sensors properly calibrated? There are statistical tests that are run on data to make sure it is trustworthy. Then the team needs to bring data together from various data streams to prepare it for specific uses in AI modeling. This requires that the data be extracted, transformed, and loaded (ETL). We cover the architecture for data management in Chapter 9.

In addition, an organization needs to establish ongoing, regular processes of data governance and quality monitoring necessary to ensure that the data is clean and to correct any gaps.

Data Labeling

The next step is to start preparing the training data. If an organization is using supervised learning, the team needs to label the data being used for training. Labels are what a company is trying to predict from data. As an example, there may be information about a customer's transactions and other activities such as the number of calls to the call center, and a field or label that says whether that customer canceled her service or not. AI models learn from this data and predict the label for new customers whose data is not in the training dataset. These labels may be inherent in the information that has been acquired: for example, whether someone defaulted on a loan. If they are not, labeling requires expert intervention, a process by which a business expert or AI scientist tags unlabeled data with meaningful information.

Consider the example of automatic detection of damage on an oil pipeline from drone images. The company has historical pictures of the pipeline, and they have been manually tagged for which ones showed damage and which ones did not. The additional labeling required would be to draw boundaries or shade the damaged parts of existing images. This precise labeling can then be used as training data.

Applying labels to data must be done accurately. Otherwise, each error reduces the predictive power of any AI model that has been trained on this dataset. If the data is not inherently labeled, it is vital to decide who does the labeling, how she will do it, and what amount of effort it will take. The type of data that needs to be labeled defines what kinds of tools are used to do the labeling. For image data – for example, to tell what parts of industrial pipes are corroded or not – there are tools such as Annotorious and LabelMe. For audio data, say, to recognize what song is playing, one could use Praat. For unstructured text or natural language, there are several open-source and commercial tools.

If the data being labeled has data privacy needs, it will likely have to be labeled in-house, or potentially by a company that works on labeling data. For less private data, such as public images to train image recognition, crowdsourcing platforms such as Amazon Mechanical Turk can be used. Outsourcing can often be less costly and faster than labeling in-house, but care must be taken so that the resulting data labels are not of low quality.

Automated labeling can also be used. One approach is to assign labels based on business rules programmatically. Another method is to use semisupervised learning, starting with only a few sets that are labeled. It is also sometimes possible to generate synthetic data, for example, using deep learning's Generative Adversarial Networks (GANs). In this approach, a deep learning algorithm generates new data to align with the statistics of the original dataset. Automated labeling and synthetic data are less expensive and usually have lower privacy concerns, but they are often less accurate than the other approaches described earlier.

Feature Engineering

With clean, labeled data available, a way for an AI scientist to begin to understand the data in preparation for AI modeling is by doing *exploratory data analysis* (EDA), such as looking for correlations and anomalies in the data. Are two variables almost entirely dependent upon each other? Are there data points that are outliers – cases in which, say, data that has been collected has values that are out of the normal range? Deciding what an outlier is can be somewhat subjective. Outliers, whether they are global outliers, contextual outliers, or collective outliers, can be the result of bad data, but they may also be legitimate. In some instances, outliers might be precisely what a business is looking for: interesting data points that would be useful for that business to understand if it is focused on, say, fraud detection. At other times, outliers get in the way. The team needs to make sure it knows which is which. Anomaly detection, while applicable in many domains, is an important part of preprocessing data in order to identify and treat outliers.

The AI scientist also looks at the summary statistics – that is, simple characterizations of the data. Percentiles can help identify the range for most of the data. Averages and medians can provide a general sense of the distribution of the data. Correlations can indicate relationships among data elements. Visualizations are often useful. An AI scientist can generate these visualizations of the data by putting it into buckets and plotting it. Significant factors to consider at this point include the mean, or average, of each dataset and its standard deviations as well as first-order correlations among the variables. Box plots help identify outliers. Histograms and density plots provide a sense of the spread of the data. Scatter plots can describe a relationship between two variables, also known as bivariate relationships. All this gives the business and the AI team a better feel for the data it has.

Once the data is well understood, the AI scientist performs feature engineering. This extracts those features that do the best job of predicting what the company wants to know. To look at customer churn, for example, a retail bank examines all the available features from a sample of its customers and figures out which features have predictive value. These might include how long an individual has been a customer, her average balance, how active her account is, the number of interactions she has had with customer support staff, even her age. The bank's training data uses all those features, as well as whether this particular customer churned (the label), to enable the bank's AI systems to learn a model that can predict what other customers are most likely to churn.

The team needs to generate and select the best features for their needs. Generated features are those that are some combination of existing data – for example, duration instead of start time and end time. There may be features generated that are useful in more than one use case; if so, the team stores and shares these features in a feature marketplace rather than re-creating them each time it is needed. The feature marketplace is explained in Chapter 9. If income is the raw data and it has been bucketed into bins of income ranges, the team makes sure that if someone else needs to use the same ranges, they are easily able to do so.

For AI modeling, not only is it essential to determine which features should be used, but these features have to be transformed to work more efficiently within a model, making models more natural to interpret, capturing more complex relationships, reducing data redundancy and dimensionality, and standardizing or normalizing variables. Feature transformations may take a variety of forms. In the context of machine learning, *normalization* means rescaling variables, so their values range between zero and one. Normalization can help calculations because a model generally uses multiple features, and normalization brings the values of various features to similar scales without overweighting one feature over another.

For instance, if the ages of a sample population range from 22 to 92, they may be mapped, with 22 mapped to 0 and 90 mapped to 1, by subtracting 22 from the age feature and dividing the result by 70 (92 minus 22). Alternatively, to make the scaling more general so it works for all known ages, age zero could be mapped to 0 and age 125 could be assigned to 1. If another feature was salary, then scaling both would prevent overweighing one versus the other based on the numbers in them. *Standardization* has a similar outcome, but it is accomplished by subtracting the average of each variable from each sample and dividing by the variable's standard deviation.

Binning and bucketing turn variables with high cardinality – that is, many possible values – to ones with low cardinality. This might include taking the entire range of those customers' ages, which is a high number – say, 18–92 – and grouping them into bins or buckets: for example, people who are 18–29, 30–39, 40–49, and so forth. These techniques can be applied to both numeric and categorical features, which means they work even if the data in question does not include any numbers. One could, for example, take all the colors of the rainbow and bin or bucket them into cool colors and warm colors, or bright colors and pastel colors. *Discretizing* again involves simplifying the values of the features a team is working with to make the calculation more straightforward, in this case by a process similar to rounding up and down. For example, the team might take people who are listed as age 18.0000, 18.9999, and everyone in between and call them all "18."

There are also ways to combine multiple features that are either dependent on each other or do a good job of predicting the same thing. Eliminating some of these can reduce redundancy in the data. If they are not removed, predictions may be less accurate because of *overindexing* on the redundant data. Overindexing occurs when one feature is excessively impacting the whole model. This process is generally called *dimensionality reduction* and is commonly done using a technique known as *principal component analysis* (PCA), in which an AI scientist looks at all the data and groups the variables so that variables in each group have the highest correlation with each other within the group, and the least correlation with the variables in other groups. The variables in each group are transformed to one feature. Now each feature is fairly independent from the other. Doing this can, however, reduce human interpretability of each feature, because it is a combination of different features. For example, through PCA, the three features of height, age, and shoe size may be combined into one.

One important aspect of feature engineering is the creation of *embeddings*. Embeddings are continuous variable vectors that represent some discrete categorical variables (e.g. customers or products). For example, instead of the category "perfume," you might use the six-dimensional vector (0.01359, 0.00075997, 0.24608, 0.2524, 1.0048, 0.06259), created from the source data using deep learning. These embeddings can then be stored in a feature marketplace and used in other classical machine learning and deep learning models as an input feature. Although this is a somewhat abstract concept, using embeddings has often improved model accuracy by as much as 10%. Often, in simpler methods such as one-hot encoding, dummy variables and label encodings are used for converting categorical to binary data types of model training and data insights.

Different models may have different feature engineering requirements or even built-in feature engineering, so transforming the input data again helps create the best inputs so that AI models can learn the most from that data.

Selecting and Training a Model

Next, the AI scientists choose the algorithm to use as the basis for the AI model. You can think of the *algorithm* as the baseline mathematical function, for example, a linear function $f(x) = ax + b$. A *model* is created by training the algorithm on a given set of data. Continuing the foregoing example, one model may be $m1(x) = 2x + 3$, or another *model* based on the same *algorithm* but using different input data could be $m2(x) = 7x + 2$. In the models, the parameters a and b from the algorithm have been trained, based on different input data.

For a predictive model to be considered good, four criteria need to be considered. First, it should have a high *model performance*. The model should be accurate – it should make useful predictions. Typically, *best* is defined as the model that has the smallest error as measured by a mathematical error function, more generally called a *loss function*. Sometimes the loss function may be a combination of lowest error and lowest bias, if fairness is an essential aspect of this model (see Chapter 10 for more on bias).

Second, it should be interpretable: it should be relatively easy to explain how the predictions were made. This may sound obvious, but it is not always the case. Deep neural nets, for example, as noted in Chapter 2, are primarily black boxes, and black boxes are not explainable. We know what goes in and what comes out, but what is going on inside is often not representable in terms that humans can understand. Decision trees, on the other hand, are generally easy to interpret in principle, but if they have thousands of branches, it may be more challenging to understand them practically. Given a company's needs, the team quantifies the advantages and disadvantages of each model.

Third, the model should be both fast and scalable. It should not take extremely long to train, and it should not take too long for it to make predictions at inference time. Models change as business requirements, data collection, and the production environment change, so a company needs a flexible model. Lastly, the model should be robust, providing consistent results when the same or similar inputs are used.

Once the team knows what it would like to predict or learn and has begun to examine the data that contains the features on which its model will be trained and will prepare training data by generating labels, team members have a decision to make. Which kind of AI or machine learning algorithm should they choose? As noted in Chapter 2, there are many to choose from, including deep neural nets, decision trees, logistic regression, and support vector machines, among others. It is essential to realize that different models, no matter how good, can be good at different things. Each of them is better at different types of tasks, although some are only small variants of others. One model might perform well for many use cases, but not all use cases. That is why an AI scientist may consider many model options, and sometimes may use more than one. When considering the options, she should be able to guide business teams through the decision of what is optimal and what trade-offs need to be made to get the results that the use case requires. Often these trade-offs will have business implications, as we will see in the example we consider in detail in Chapter 13.

As an AI scientist starts the modeling process, she needs to carefully consider her approach for sampling the data for training, testing, and validating models. Data sampling is the statistical method for selecting observations (rows) from the dataset to be able to estimate parameters about the population. It feels simple but can often be tricky. Sampling needs to be done keeping in mind the distribution, size and type of data, and the available computational power. Broadly speaking, there are a few major types depending on the use case and algorithm being used: random sampling, stratified sampling, cluster sampling, multistage sampling, and systematic sampling.

To create the best learning model, the AI scientist generally splits the data into two groups (see Figure 8.4). About 60–80% of the data is used for training the model, and the remaining data is used to test the model. As discussed in Chapter 2, the purpose of machine learning is to accurately predict based on data that has

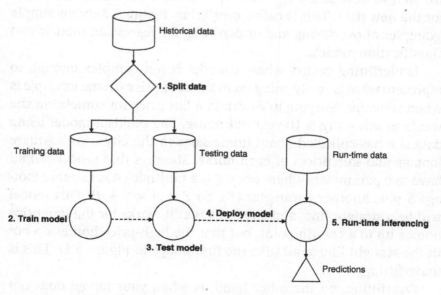

Figure 8.4 Process for training and validating the model.

not been seen before by the model. That is why the data is split into two groups – the first set for training and the second for use as the "previously unseen" data on which prediction accuracy is measured. This tells us whether the model is generalizing sufficiently to be predictive with unseen data, rather than just being descriptive of prior data.

A few concepts are essential to understanding model performance or model accuracy. First, a model is defined by a set of model *parameters*. These parameters are those variables in a model that are learned through the training process. The training process explores different combinations of model parameters and selects the ones that give the lowest error (or optimal loss). The more parameters a model has, the more complex it is – that is, the more parameters it must learn. This is sometimes referred to as *model complexity* or *model capacity*. If you have too few parameters for the training data, you do not get a very accurate model. This is called *underfitting*. On the other hand, if you have too many parameters, your model starts to become very specific to the training data and does not generalize well to new datasets – that is, it is not a very accurate model for the new data. This is called *overfitting*. Figure 8.5 shows simple examples of overfitting and underfitting for regression models and classification models.

Underfitting occurs when a model is not complex enough to represent what is really going on in the data. An extreme example is when someone is trying to ascertain a fair price for something she wants to sell – say, a 10-year-old house. She builds a model using data she has collected about home sales in the city – with square footage and sale prices of each home. She says that model should have two parameters where price P is a multiple (A) of square footage S plus another parameter (B). So $P = A \times S + B$. This model will be a straight line, and she finds that it works for the low-price homes up to a certain point, but that the high-price homes do not fit the straight line at all (like the first image in Figure 8.5). This is underfitting.

Overfitting, on the other hand, is when your model does not perform well on data other than your training data. Continuing our

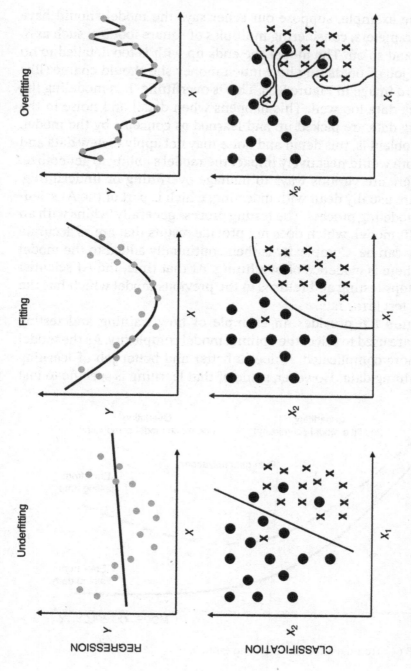

Figure 8.5 Underfitting and overfitting for regression models (top) and for classification models (bottom).

extreme example, suppose our seller says the model should have 100 parameters, considering multiples of square footage, such as S, S^2, S^3, and so on. The model she ends up with is too detailed to do a good job of predicting how much money she should charge (like the third image in Figure 8.5). This is overfitting. It is modeling the training data too well. This happens when detail and noise in the training data are picked up and learned as concepts by the model. The problem is, this detail and noise may not apply to new data and therefore would negatively impact the model's ability to generalize.

There are various ways to manage overfitting or underfitting. They are usually dealt with in testing, which is part of the AI scientist's modeling process. The testing process generally begins with an underfit model, which does not provide results that are as accurate as they can be. Complexity is then continually added to the model until there is evidence of overfitting. At that time, the AI scientist then stops testing and returns to the previous model which has the lowest test error rate.

Figure 8.6 provides an example of how training and testing errors are used to select the optimal model complexity. As the model gets more complicated, it does a better and better job of learning the training data. However, some of that learning is specific to just

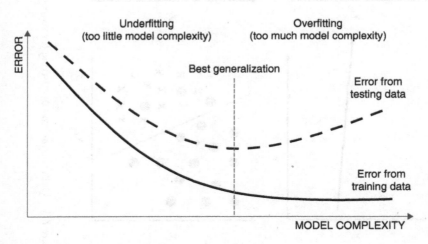

Figure 8.6   Training error versus testing error.

the data it was trained on and does not represent the characteristics of the data in general. The problem arises after training, when the model is provided with data that it has never seen before. The model may well try to explain idiosyncrasies present in the training process by making predictions that are not as accurate as they could be. In this situation it probably is not going to be as successful as it might be in predicting from new data. The model that is most generalized for unseen data is where the test set error is lowest, shown as "best generalization" in Figure 8.6.

As mentioned earlier, data is split so that part of it can be used for training and part for testing. Breaking the data allows testing a model on different data than it was trained on. Typically, models are trained on approximately 60% to 80% of the data, leaving the rest for testing. Another approach that is commonly used is called k-*fold cross validation*. This occurs when the data is split into k groups (say, $k = 5$) and then the first four groups are used for training and the last for testing. Next, groups 1, 2, 3, and 5 are used for training, and group 4 is used for testing, and so on. (Note that if the data is a time series, this approach to k-fold will not work.) The average error rate of the k runs is used as the model's error rate. Therefore, k-fold cross validation improves the model performance and generalization and is often used when the total dataset size is small. Other popular methods are stratified k-fold cross validation and leave-one-out cross validation.

In machine learning classification problems, there is an additional step needed to test and tune the model's performance. This is *hyperparameter* optimization. A hyperparameter is a parameter that is external to a model, set by the AI scientist before the training process starts and not learned through the model training process. Often the output of classification models is a probability for what the class could be. The AI scientist can then define a hyperparameter that sets the cutoff above which the class would be predicted to be "Yes/Positive" and below which the class would be predicted to be "No/Negative." For example, in a "does the patient have cancer" model, this cutoff could initially be set at 50%. The AI scientist can then look not just at the accuracy of the model but also at the false positives and false negatives it gives and adjust the 50% cutoff up or down to improve the model's prediction.

One way to do this is by using a *confusion matrix*, which has the counts of the class predicted by the model by row and the actual counts of the class by column (see Figure 8.7). In our example, we have two categories – positive and negative for cancer diagnosis – but the same applies to more classes; you would, for example, have a 5 × 5 confusion matrix for a problem with five categories. The four quadrants of the two-class matrix are:

1. True positives (TP) – count of patients with cancer and predicted accurately to have cancer.
2. True negatives (TN) – count of patients without cancer and predicted accurately to not have cancer.
3. False positives (FP) – count of patients without cancer but predicted incorrectly to have cancer (also called Type 1 error).
4. False negatives (FN) – count of patients with cancer but predicted incorrectly to not have cancer (also called Type 2 error).

In this example, a false negative may have potentially life-threatening consequences, whereas a false positive is not dire, although it will likely provide the patient with a bad experience and incur unnecessary costs. It is therefore desirable to reduce false-negative rates while improving true-positive and true-negative

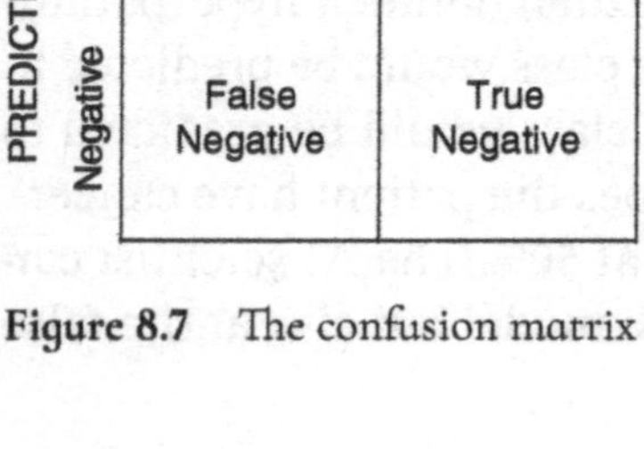

	ACTUAL VALUE	
	Positive	Negative
PREDICTED VALUE — Positive	True Positive	False Positive
PREDICTED VALUE — Negative	False Negative	True Negative

Figure 8.7   The confusion matrix setup.

rates. Two metrics often used are the true-positive rate (TPR), which measures the percentage of true positives (TP) out of everything predicted to be positive (TP + FN); and the false-positive rate (FPR), which measures the percentage of false positives (FP) out of everything that is actually negative (FP + TN).

The receiver operating characteristics (ROC) curve in Figure 8.8 shows the true-positive rate against the false-positive rate of the model. Ideally, TPR should be as high as possible, and FPR should be as low as possible. For this to happen, the area under the curve (AUC) needs to be as large as possible, hence pushing the curve to the upper left. The modeler selects a point on the curve that is closest to the top left of the axes, thereby choosing a balance between false positives and false negatives.

Through this process, the AI scientist picks a model after experimenting with multiple algorithms and addressing any underfitting or overfitting problems. Model selection ensures that the ROC curve is as far to the left and top as possible and the area under the curve is the largest it can be. This indicates the model is as good as it will get. The AI scientist then tunes the hyperparameter, selecting

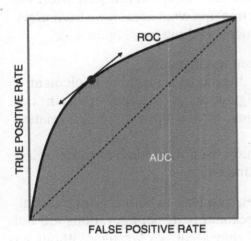

Figure 8.8 Receiver operating characteristics (ROC) curve and the area under the curve (AUC).

the optimal location on this specific ROC curve for solving the business problem (shown by the arrow in Figure 8.8). For the highest accuracy, irrespective of false-negative and false-positive rates, it is best to be at the point on the ROC curve that is closest to the top left corner. In the cancer diagnosis example, the AI scientist wants to reduce false-negative rates and trade off against false positives if needed, so she sets the hyperparameter such that the cutoff threshold (hyperparameter) sits further to the right on the ROC curve.

Managing Models

Building real-world machine learning algorithms is complex and highly iterative. An AI scientist may build tens or even hundreds of models before arriving at one that best meets the acceptance criteria. Keeping track of those models is a little like running a search mission for some hikers lost in the Himalayan foothills. The areas that rescuers search, even areas where the hikers are not found, must be clearly marked. Otherwise, searchers waste precious time repeatedly combing through the same areas. Similarly, model management is essential to cover all the configurations that affect the performance of the model and tracking what areas have been previously covered or not. This includes tracking (see Figure 8.9):

- What data was used for training the model
- What use case or variation this model is meant to implement
- What infrastructure configuration is required to train this model or use during inference, as well as any other information required for deployment
- What model experiments have been tried, such as model selection or hyperparameter optimization

Not keeping track of models can lead to both accruing technical debt and losing model repeatability. If configurations are not tracked properly, a variety of things can go wrong. Without an organized record of the experiments performed, it is impossible to ascertain whether someone has already tried something identical

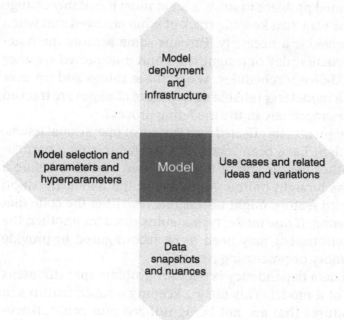

Figure 8.9 Comprehensive model management spans four types of configurations.

or very similar to what someone else is about to do. Not tracking experiments is not only inefficient and costly; it makes collaborating, reviewing, revising, and building on others' work more difficult. Without a process in place, it is harder to do meta-analyses across models, and the team loses time backtracking when no one can easily find out, for example, which parameters were used for an earlier result. It is also harder to track model versions, such as what changed between model version 233 and version 237.

A potential problem can arise when adding, changing, or removing features in an AI model as development moves forward. Changes generally require that the model be revalidated, that is, run through learning and testing again, since configurations that have changed may affect the validity of predictions. When there is a very complicated problem to solve, a team must inevitably change things along the way. Just keeping track of what changed and when and how it evolved is a necessity. Perhaps some sensors malfunctioned on a particular day or a significant and unexpected weather event affected delivery schedules. When these things add up over time, it makes a model less reliable unless these changes are tracked and handled appropriately in the modeling process.

Features might also be affected by a variety of operational issues. One feature might have been incorrectly logged during a particular week. Another feature might not have been available before a new sensor was brought online. A change in the data acquisition format for a third feature might necessitate rewriting the code that does preprocessing. If one model type is substituted for another, the computing environment may need to be reconfigured to provide additional memory or processing power.

Unneeded data dependency is another problem that threatens the reliability of a model. This means keeping unused features in the model: features that are not being utilized and could, therefore, be removed. This often happens because there was no time to take them out, or a team was not organized enough to realize they were no longer being used. That is fine if these features never affect outcomes, but if the data changes, they can adversely impact results.

Another potential problem is a pipeline jungle. This happens when there is no clear data architecture or well-defined standards to make all the sequential elements in processes coherent end to end so that the data can flow seamlessly in the production process. If there is a pipeline jungle, the DevOps team finds that data preparation lacks consistency over time, making it hard to track what depends on what. This means that things such as detecting errors and recovering from failures become very complex, making production inefficient or impractical.

As an organization explores opportunities to use AI, it needs a formal approach for keeping track of things: testing the experiments, capturing what works, and maintaining an "idea graveyard" for concepts that have been tested and determined to be untenable. A process like this makes things easier for both individuals and the organization. It allows AI scientists to track their work in detail, giving them a record of their experiments. It also enables them to capture meaningful insights along the way, from how normalization affected results to how granular features appear to affect performance for a specific subset of data. No organization achieves the necessary level of visibility and analysis when managing models via spreadsheet. A far more sophisticated system is needed to document what has been done and what the results are. To succeed at an enterprise scale, an organization must be able to store, track, version, and index models as well as data pipelines. This enables models to be queried, reproduced, analyzed, and shared, even if those models are not the ones eventually used. Sound model management empowers AI scientists to review, revise, and build on each other's work, helping accelerate progress and avoid wasted time.

Also, the team may lose insights and sacrifice opportunities to make sense of what it is seeing. These might include how individual parameters that are contributing to model complexity have affected results, or the ability to notice that a particular feature seems to improve the performance for a specific subset of data. The potential quantity of ideas and nuances can quickly become overwhelming. To avoid that complexity, teams should design and implement an automated model management process for tracking and managing the lifecycle of use cases, model management, and experimentation. Doing so helps in tracking idea performance and ensuring the quality of those ideas. It is also far more efficient to provide a team with information about and easy access to successful and unsuccessful ideas to avoid duplicate efforts and potential conflicts. It also enables the organization to conduct meta-analyses across models to answer broader questions, such as what hyperparameter settings work best for different types of models or features. All this accelerates the efficiency of the deployment of AI services.

Testing, Deploying, and Activating Models

Even after an AI scientist comes up with a model that scales well, there is quite a lot that goes into deploying an AI project that is not handled in earlier stages. The model has to be tested. A model risk governance team must issue final approvals. The model must be integrated with the applications that use it. Applications must be extended to get the results from the models in front of people who need them, and a mechanism must be in place for those users to provide feedback.

Testing

Continually testing is essential for creating and running robust models in reasonable amounts of time. As discussed earlier, during its creation, the AI model must be verified for model accuracy, which needs to be at a certain threshold to be practical for the use case. After the model is built and integrated into the data and applications that use it, it must go through other forms of testing, just like any new IT application or technology module being deployed. One of these is end-to-end functional testing. Does the model work in a production-like environment, able to receive the data for inference? Does it produce outputs or predictions at the expected level of accuracy? Do the consuming applications receive the outputs and either display or act on them?

The end-to-end model also needs to go through performance testing. Does it return results promptly, with the expected throughput given the expected load of data? In some cases, the code written by the AI scientist may not be production scale, and it may have to be rewritten for higher scalability. Another test the team may need to execute is for model cascades. A model cascade is when one model feeds outputs into a subsequent model, which uses these outputs as its inputs. In this situation, care must be taken with the way data is handled. What happens if the outputs from the second model are incorrect? How can the team ascertain where the problem arose? Was it within the first model or the second one?

Testing, such as integration testing, user acceptance testing, staging, and finally, production testing, needs to continue as a model is progressed through the different environments. Some additional types of testing are sometimes conducted in the production environment. One is called a *canary test* or canary deployment. Canary testing is when the deployed model is only exposed to a small set of users (usually less than 5%) and monitored to ensure it is working well. The test acts as the proverbial canary in the coal mine. If it is successful, the model can be deployed to all users. The other production test is when two or more models are deployed to see which one may work best in the real world. A/B testing deploys different versions of the models in parallel, and different sets of users are configured to use different models. A/B testing is often used when a model is being enhanced or upgraded. The old and new models are A/B tested, and the better-performing model is the one left in place. In some cases, A/B testing is used to test hypotheses where A is the test group and B is the control group.

Governing Model Risk

Teams should utilize model risk governance to reduce any risks resulting from the use of an AI model. Model governance applies before the model is developed, while it is being developed, and afterward. Before modeling, the focus is on data regarding quality and bias. During modeling, the focus is on model bias and interpretability, in addition to model accuracy. After modeling, the team executes compliance testing, such as sensitivity analysis and bias testing. Also, a fail-safe mechanism must be put in place in case something goes wrong with the model. Utilizing industry best practices such as these ensures that an enterprise, and its model risk governance team, has done the best job it can to implement safe AI. (For a complete review of model risk governance, see Chapter 10.)

Deploying the Model

Once a model has passed testing and is approved by the governance process, the team must make it easy to deploy in each environment. This means writing scripts to automate the deployment process. It is better to be able to push models to environments with one-click deployment rather than having to input manual commands each time. All these activities should be automated using the same types of tools being leveraged across software engineering practices. Once an AI scientist has helped to decide on the model or models that have been tested, she needs the support of an experienced AI DevOps or machine learning engineering team. The AI DevOps team is as important as the AI scientist herself: it is responsible for setting up a continuous integration/continuous deployment (CI/CD) pipeline for AI models.

After the AI scientist has checked in the code for the AI model, AI engineers package the AI model and feature engineering microservices and create a model deployment unit as shown in Figure 8.10. This model deployment unit is what is deployed to the test and production environments as needed, including the required infrastructure configuration changes. Next, the team installs the model deployment unit and automatically executes the relevant tests mentioned earlier. It sends notifications for remediation in case of any failures, or if this is not necessary, it tears down the environment and progresses to the next stage.

Activating the Model

Once the model has been validated and deployed, developers write code in the enterprise application that will use the output of the AI model. This code is used to invoke, or activate, the model every time a prediction needs to be made. There are many ways to do this. For example, if the machine learning algorithm is trying to decide if an email is spam or not, the email system passes the email to the model through an *application program interface* (API) each time a new email is received. An API is software that can accept input

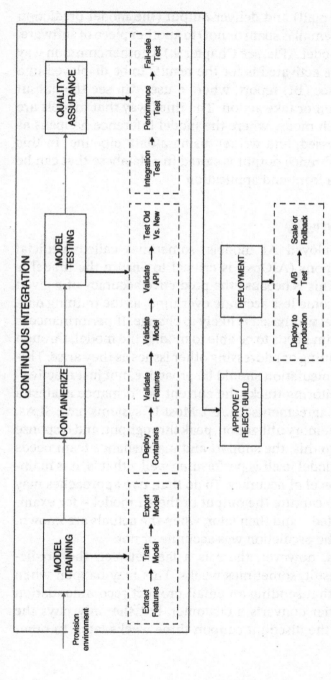

Figure 8.10 AI DevOps process.

167

(in this case, the email) and deliver output (the model prediction about whether the email is spam or not) to another piece of software. (For more about model APIs, see Chapter 9.) Another common way for the model to be activated is for the results to be displayed in a business intelligence (BI) report, where a user can see the output and make a decision or take action. The third way that models are deployed is in batch mode, where the model inference happens as data is read, processed, and written using a data pipeline. In this case, the model inference output is stored in a database that can be accessed through a front-end application.

Production Monitoring

Monitoring of deployed AI models, sometimes called artificial intelligence operations (AIOps), is critical to ensure the model is performing well. This is because the predictive accuracy of a given model tends to become less accurate over time, as the training data on which the model was based is likely to change. If performance is inadequate, the team needs to be able to modify the model, through either model retraining or addressing other issues as they arise. This monitoring and remediation should be proactive, not just reactive.

Proactive monitoring tracks the current performance against a set of service level agreements (SLAs). Most IT systems have SLAs, such as CPU and memory utilization, peak throughput, and response time. In addition to this, the support and maintenance team needs to know that the model itself is performing well – that is, it is maintaining the same level of accuracy. To do that, two approaches may be taken. One is to capture the output of the AI model – for example, what it predicted – and then later, when the actuals are known, capture whether the prediction was accurate or not.

In some cases, however, there is a lag between the prediction and actual result, sometimes weeks. This may happen when a model predicts that sending an email product recommendation and a discount offer converts a customer, but she only buys the product and uses the discount coupon three weeks later. In other cases, the actuals may never be known because some action was taken to mitigate the prediction: for example, the customer who was predicted to churn never left because she was given a discount based on the churn prediction. In these cases, causal inference needs to be modeled to connect the dots. The other approach is to monitor *feature drift*, a condition in which the statistics of input data start to change. An example of feature drift is an AI model that predicts the likelihood of default on a loan for a customer in which one of the input features is an income distribution of the customers. If that distribution materially changes, with more customers now in higher or lower income ranges than in the training data, this is considered feature drift. This may be an indication that the learning from the training dataset may be becoming stale. Feature drift is a leading indicator that the model needs a refresh, beyond some threshold. Tracking prediction against actuals is a lagging indicator.

Conclusion

This chapter has described the process that AI teams go through to implement an AI model as part of a use case. However, the process does not stand on its own – it needs people to be organized (see Chapter 11) including roles that are required in most IT projects, and it requires a platform (see Chapter 9) to ease and speed up the processes. The process described in this chapter can be viewed as the requirements for the details covered in the rest of Part III of this book. In Part IV, we describe a working example and create a model using this process for readers interested in more detail. All these – the process, the organization, the platform – work in concert to enable the success of enterprise AI ventures.

Chapter 9
Building the Perfect AI Engine

> *How hard is it to build an intelligent machine? I don't think it's so hard, but that's my opinion, and I've written two books on how I think one should do it. The basic idea I promote is that you mustn't look for a magic bullet. You mustn't look for one wonderful way to solve all problems. Instead you want to look for 20 or 30 ways to solve different kinds of problems. And to build some kind of higher administrative device that figures out what kind of problem you have and what method to use.*
>
> Marvin Minsky, professor of artificial intelligence at MIT

There is a long-held belief that the most important part of building an AI initiative is designing the AI application – the algorithm or software model created by AI scientists to make predictions. But as we have already seen, most of the effort that goes into building an enterprise AI application is spent on data gathering, cleansing, and labeling; creating data pipelines; DevOps; deployment; building business applications for end users; and monitoring and enhancing it over time. When companies ignore these stages, they often end up with highly inefficient modeling processes and people working in individual silos. This wastes money and time and delays getting business benefits from the model. The way to avoid this is to build a holistic AI platform.

AI Platforms versus AI Applications

An AI platform is a cohesive, well-integrated software solution running on scalable hardware that accelerates the full lifecycle of AI projects and applications across an enterprise. AI platforms lighten the workload, encourage cooperation, and speed up adoption. Robust platforms support all tasks across the whole AI lifecycle. Such a platform is an essential foundation of a company's AI applications. AI applications are built and run on AI platforms; they consist of the software that solves specific problems or answers particular questions. In the last 10 years, companies that have used AI have focused predominantly on building applications. For example, there are retail companies that built one AI application for targeting marketing messages, another for online recommendations, and yet another for demand forecasting in the supply chain. Each of these applications needed its own version of data management, computational infrastructure, deployment processes, and AI models.

Today, leading companies build AI platforms on which to develop and implement these applications. This platform mindset is partly driven by the fact that it costs less to build the platform once and then build applications on it, and partly because it allows more data to be shared across departments such as marketing, e-commerce, and supply chain.

What AI Platform Architectures Should Do

Stated broadly, the essential things the AI platform should do are enable an AI enterprise to grow and improve over time as technology and needs evolve, and be easy to use and therefore available to more of a company's staff – and do this at enterprise scale. More specifically, an AI platform should increase the probability of success by enabling the team to run a higher number of experiments without growing the costs linearly. The platform should provide self-service access to the foundational AI technology to reduce friction from AI teams, enabling their productivity, speed, and scale.

Increasing the number of experiments was the insight the Wright brothers had that helped them to fly before anyone else. The Wrights' competitors would take nine months to build an airplane and then would try to fly it by throwing it off a (small) cliff. After it crashed, they would pick up the pieces, go back to their workshop and analyze them, and try to figure out what to improve. When the Wrights recognized this, they asked an important question: How can we increase the number of experiments we do? To do so, they built the world's first wind tunnel. Now, instead of testing a plane design every nine months, they could build a new model every day, put it in the wind tunnel, and learn from it, doing 270 experiments at lower risk and lower cost than their competitors in the amount of time their competitors took to do one.

Like the Wright brothers, AI scientists need to spend their time in rapid experimentation to learn faster, without having to wait or waste time when they provision an environment, deal with data ingestion and cleansing, or manage computing power. Each AI scientist must be able to quickly tap into the tools she needs, when she needs it. Ensuring that each of these users' experiences is as simple and efficient as possible requires thoughtful consideration regarding what components to include in a platform.

A flexible AI platform enables the building of multiple applications within the same technical framework. This flexibility allows future products and applications to be developed and deployed more quickly and cost-effectively than if they were built as standalone applications. For example, at a bank, an AI platform can support all functions across different divisions and enable the use of shared components. The platform can be used to develop and deploy models for anti–money laundering (AML), personalization in customer servicing, and credit approvals. The AML models integrate into the financial-crime-prevention workflow in the AML application; the personalization models integrate into the bank's website and mobile app; and the credit models for predicting loan default risk integrate into the credit workflow in the credit application. Over time, a well-built platform can even enable a company to expand its AI network to include suppliers, vendors, and partners, allowing them to build on the platform as well. The company can reap many benefits from this expansion of its AI network.

An AI platform needs to fit seamlessly into an enterprise. It has to support analysts, product developers, and other stakeholders, as well as both AI models built in-house, and commercially developed models that have been purchased. It also has to enable a deployment team to easily access and configure the data, AI models, and computing resources in a simple, efficient, and reliable manner. To do this means the platform needs to include automated tools that are easy to use and minimize repetitive tasks.

With the fast pace of development in the field, new capabilities and services are continually becoming available, including new algorithms, open-source code, components and solutions from startups, and new AI frameworks. At the same time, it is difficult if not impossible to predict any company's future AI needs. The ideal AI platform, therefore, is able to evolve. Not limited to any one AI type or solution, it makes possible rapid provisioning of a high-performance computing environment to support multiple kinds of AI. In short, it transforms AI from a series of finite point-solutions to an enterprise capability that can be improved over time to meet a company's evolving business needs. When a platform can do this, it is said to have an *evolutionary architecture*.

To help this evolution, an AI platform should include a portal where users can connect and share information about what they are doing as well as which tools and features they are using or know about that might be helpful to colleagues. The portal should be a place where AI scientists can request that new open-source or commercial software be included in the platform, or where users can capture ideas for new use cases. It should also be where governance or procurement workflows are initiated. A portal facilitates faster, more efficient, and more effective collaboration among AI scientists and AI engineers, enabling a more even distribution of work and allowing completion of research and experiments efficiently and on schedule.

There are those who may feel that tackling the complexity of building an AI platform is a lot like building an airplane while in flight. They are concerned that new technologies are emerging faster than their budget cycles. When an executive cannot be sure of what she will need in the future or what solutions meet those needs, how can she decide where to invest? The answer is to focus not on the technologies but on the overall architecture. Utilizing an evolutionary architecture makes it faster and less costly to replace technologies either as they become obsolete or as better options emerge. The most successful organizations at AI are those that take the time to assemble an appropriate enterprise AI platform for their businesses, enabling them to deliver more value more quickly- not just today but as new opportunities take shape in the future. The investment that companies make in an AI platform should be adequate to accomplish all of this.

A good AI platform enables a business to quickly increase productivity. Using key performance indicators (KPIs) and comparing results before and after implementing a robust AI platform, companies can easily see the benefits of the platform approach. Some companies witness the same AI teams doing up to five times as many projects per year on a well-designed platform than they were able to before they had one. Other companies increase the speed of building new applications on their AI platform from an average of eight months per project to eight weeks – with the same team size. Some companies scale to 30 times as many experiments in the same time frame using a platform, compared to what they did before. An AI platform also enables the building of large AI teams without being slowed down. In doing all this, the platform helps in tackling skills gaps, serving as a focal point for onboarding new talent, and speeding up the ramp-up of new team members as well as distributing work more effectively among them. It can also help support best practices throughout a team of AI scientists.

Many AI projects are abandoned because they take too long to accomplish. Often, the slowest part of the process is provisioning the data environment and the high-performance computation environments by IT infrastructure teams. Other delays that may cause abandonment may come from ad hoc or manual model risk governance and assurance processes that take too much time to execute and be approved. Delays can also be caused by a lack of rigor, structure, and automation in deploying and hosting production models.

Figure 9.1 tells part of the story of what a good AI platform can do. Companies at a low stage of AI platform maturity are on the solid line; these companies must continually add data scientists to put increasing numbers of AI models in production. Companies on the spaced-dotted line have put scalable AI platforms into place. Even businesses that must utilize rigorous model risk governance because they are subject to high regulatory pressure find themselves on the short-dotted line and not the solid line if they have the right evolutionary architectures in place.

Figure 9.2 shows where some of the speed and scale of a good AI platform come from, as well as their relative impacts. Benefits are derived from integrating existing models or components into the platform and building reusable components through each project; these components can be added to the platform and be reused on future projects to shorten their timelines. There are also benefits to automating the AI scientists' work by integrating incremental tools such as ones for autogenerating and assessing

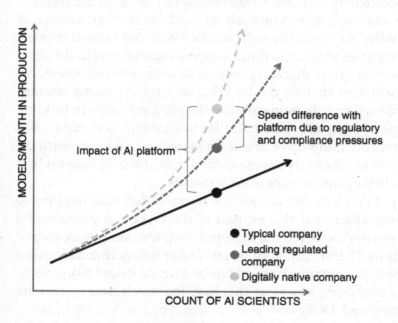

Figure 9.1   Impact of using an AI platform.

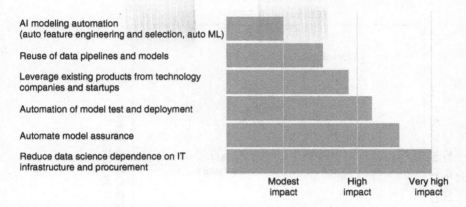

Figure 9.2   Summary of benefits of using an AI platform.

features, autogenerating models using tools such as AutoML, and data synthesis and data labeling workflows. AI scientists should be able to spend their time in rapid experimentation without having to battle environments provisioning, data ingestion and cleansing, and limited computing power, while working to easily and quickly take their most successful models live. Employees must be able to quickly tap into the tool or combination of tools they need.

Ultimately, the needs of its "users" should determine what kind of AI platform to build, paving the way for an organization to deliver intelligent services and products, both now and in the future. And by users, were not just talking about AI scientists or even company employees. We mean anyone who interacts with the platform (see Figure 9.3), from AI scientists to IT departments to software engineers to strategists, customers, and beyond. Technical users include AI scientists who create new models, DevOps engineers who deploy models, and developers who integrate models into other applications. Business users include business analysts who use existing models for analysis and end users operating the business who use the model output. Each will utilize the platform in different ways, and the platform should be able to engage with all of them in their contexts, exchanging information in optimal ways.

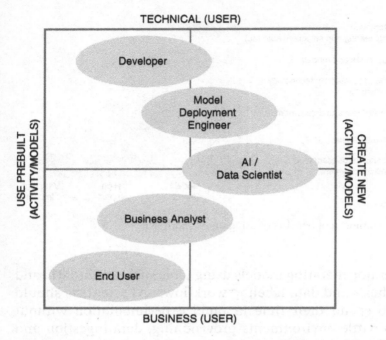

Figure 9.3 Types of users of an AI platform (vertical axis) and how they engage with AI models (horizontal axis).

Users should not have to build new data pipelines by hand, provision new infrastructure, or manually adapt the model for production systems; machine learning platforms streamline both development and production environments, make infrastructure management simple enough for self-service, and scale resources up or down when required. Interactions should be intuitive, requiring minimal user training. Some users may prefer immersive visual responses, or to engage with the system on their mobile or voice devices; these should be available when needed. Other essential user tools include mechanisms to manage containerized deployment of the various AI models across the platform and dashboards that monitor the performance of the platform.

Some Important Considerations

Before building an AI platform, an enterprise architect should focus on a few critical considerations.

Should a System Be Cloud-Enabled, Onsite at an Organization, or a Hybrid of the Two?

There are pros and cons to each. The cloud excels in many areas. Using it gets businesses to market faster: there are fewer startup and maintenance costs using the cloud, and disaster recovery can be quicker and less expensive, given the number of cloud data centers and their geographic diversity. Regarding productivity, the cloud enables companies to outsource some IT functions to specialized experts whom it may be difficult for a company to hire, train, and retain. Cloud usage means avoiding the task of coordinating different onsite teams as well as potentially cumbersome and bureaucratic management processes. Using the cloud also means not having to build and manage an infrastructure before beginning to develop models that create value. In the cloud, a simple software or operator interaction gets the enterprise what it needs; on-premise, teams might have to deal with processes that involve a web of approvals and delays. The cloud also allows for a great deal of scalability, elasticity, and flexibility. Because of these features, even people who build infrastructures in-house plan cloud migrations when they are at scale.

There are areas, however, in which the cloud may come up short. In the cloud, companies pay only for what they use, and software-licensing fees are included. However, those companies may be paying more to get benefits such as those lower costs and rapid disaster recovery. If they are fully utilizing their onsite data centers and roughly the optimal amount of their software, processing, and networking capabilities, they may end up paying less than they would if they outsourced to the cloud. Data security and privacy are other issues to consider. Many organizations believe that their data is safer on their premises than it would be in the cloud. An executive may find herself in a lengthy debate with her enterprise architect and chief information security officer on this point, and for regulated businesses, data compliance has to be planned, tested, and approved before the cloud can be used.

Latency should also be kept in mind. Despite the ease and speed of working in the cloud, information cannot travel faster than the speed of light. If a company is performing functions such as high-frequency stock trading, the time it takes for data to get from a model to a cloud data center and back can present a big problem. Traders put their platforms near network hubs so they do not lose time for precisely this reason. The flexibility, performance, and scalability required in modern AI platforms may mean a primarily cloud-based solution is best suited to a given business. There is also the option of a hybrid, using local storage for areas of an enterprise in which there are advantages to being on premises – for example, in situations involving data security or low latency. With managed software such as OpenShift from RedHat and Anthos from Google, managing multicloud and cloud-and-on-premises environments is becoming easier.

Should a Business Store Its Data in a Data Warehouse, a Data Lake, or a Data Marketplace?

Both data warehouses and data lakes are comprised of processors and the software that runs on them and storage devices and the networking that connects them. The difference between them is the structure of the data each can store. A data warehouse stores highly structured data in predesignated formats to optimize it for standard reporting. Structured data has well-defined fields, such as name, address, phone number, and so forth, and rows of data for these fields. Unstructured data is text that might be acquired from sources such as emails or tweets, or images and videos such as X-rays or MRI scans.

Data warehouses work best when a company's information and data requirements evolve slowly. There are few businesses today, however, that are not in a constant state of flux. Given the constantly changing business climate, rapidly evolving technical landscape, and the relentless accumulation of data, data warehouses alone might not serve a company's needs. Traditional data warehouses bind collected data to highly structured and rigid categories in advance; data lakes maintain raw data in its original format: for example, text files. In other words, data lakes require less transformation of the data before storing it (although it will require some during retrieval). A data lake can accommodate new types of data as the needs of an organization evolve, essentially creating a pool of information that can be accessed and queried at any time. It can also aggregate and standardize a wide variety of data as well as enable single longitudinal views of data.

A data marketplace is a data lake with an organized, accessible layer to make data discovery and business use more natural for non-IT teams. This is extremely important because it enables businesspeople and AI scientists to begin to speak the same data language. Data marketplaces make it easier for business users to find and understand the data they need, enabling them to obtain data without involving programming or IT. A data marketplace also has systems and tools in place that limit data access to authorized users. Data marketplaces help to transform data from a technical asset to a business asset, driving the company's cultural change toward greater data literacy and more data-driven decisions. This cultural change helps improve sustained data quality as well because it encourages more data usage. Business users using data marketplaces get more comfortable asking for and using more sophisticated models over time, creating greater adoption of AI.

Because of these differences, enterprise architects are moving toward data lake storage and data marketplace usage, although most enterprises currently also use data warehouses if needs dictate. The decision to use one or the other is based on a company's use cases. As discussed in Chapter 7, for companies focusing just on standardization, control, and reporting, data warehouses may be sufficient. For companies looking for flexibility, sharing, and different types of uses, data lakes and marketplaces are becoming critical.

Should a Business Use Batch or Real-Time Processing?

There are three ways to do processing for a business: batch, real-time (or streaming), or a hybrid of the two. The data collection could be batch or streaming. It can cause confusion to say a model is batch or real time if we do not specify what the processing is – whether we are referring to model training or model inference. For AI models, each of these activities could be batch or real-time (see Figure 9.4). In most applications, the training happens in batch mode, and the inferencing happens in either batch or real-time modes. With reinforcement learning and similar methods becoming more common, real-time training is being used more often. Finally, as AI models are improving, many decisions are being acted on autonomously without human intervention. Some people refer to this autonomous decision-making as real time. Whether real-time processing is required, and if so for what type, will depend on the use cases that the business is trying to solve.

Batch processing, also called batch inference, refers to inputs that are accumulated over a certain set time and then processed on a recurring schedule (say, hourly or daily). Resulting predictions are stored in a database that can be accessed by developers or end users. Batch inference can be simple, efficient, and cost-effective, which is

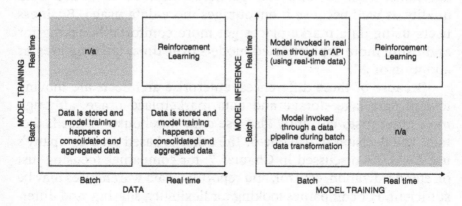

Figure 9.4   Batch versus real time for data, model training, and model inferencing.

why it is commonly used. It works well with large volumes of data and in situations in which it is acceptable to update the AI model results only periodically. For example, a business doing a valuation on homes based on size, number of bedrooms, and location does not necessarily need to update the value of each home throughout the day.

Batch inferencing, however, may not satisfy an enterprise if it needs either time-critical decision-making or interactive applications. Figure 9.5 shows the different scenarios for batch and real-time processing. Real-time inferencing is required when the data is just becoming available and the user needs to see it at the moment (autonomous vehicles, for instance, require real-time inference). But it is more common for businesses to use near-real-time inferencing. For example, a recommendation engine that recommends items based on someone's online shopping cart must take real-time data about the shopping cart and run the AI model to get

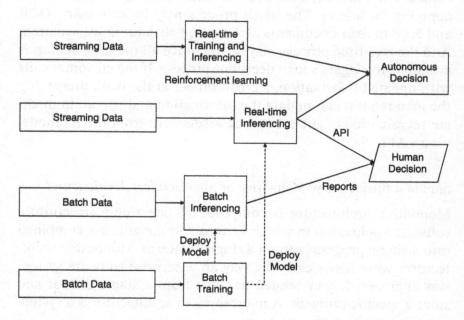

Figure 9.5 The different patterns of batch or streaming data, model training, model inference, and usage.

the recommendations. (If the recommendations were based only on what the most popular items were, the AI model could be run in batch mode and the results stored.) Inference in near real time delivers results on demand using the latest data, usually invoked through an API call. Using an API enables the delivery of output from an AI model in near real time or on demand to another program, where the output is integrated into the second application's workflow, screen, or report.

As AI models improve, many decisions are being acted on autonomously without human intervention. Some people refer to this as real-time, or streaming, decision-making. Whether real-time decision-making is required depends on the use case; most systems in which decisions have been automated are using this kind of real-time inferencing.

In the hybrid form of processing, a business can batch inference-specific outcomes and use near-real-time inference for other outcomes. For example, consider the example of a bank customer applying for a loan. The batch process may include using OCR and NLP to read documents and capture structured information. And the real-time process may be that once all the information is available, it triggers a loan decisions process. If the customer calls with updated information, a credit officer at the bank processing the loan request can update the information and run an immediate reevaluation of the credit risk assessment from the AI model via an API.

Should a Business Use Monolithic or Microservices Architecture?

Monolithic architecture is composed of one single structure: a software application in which various components are combined into a single program run on a single platform. Monolithic architectures were historically the way an enterprise software system was engineered; they tended to come from a single vendor and solve a specific problem. A microservices architecture is a system composed of numerous, loosely coupled microservices. These services are distinct, isolated, and reusable, and they work together to provide a single business capability. Today, it is becoming more common to incorporate specific guidelines drawn from the world of microservices and event-driven architectures into a system, because these architectures can be modified depending on a company's needs, making it easy to create or retire a single microservice without affecting any other microservice. This feature, known as *isolation*, provides the ability to evolve a platform incrementally as new services are developed or become available, without the need to re-architect from the ground up. It reduces time to market and time to value and enables the evolutionary structure of AI platforms.

Software that is microservices-oriented is more straightforward to scale, reuse, and integrate both within and across enterprises. In AI systems, where flexibility and rapid adaptation is critical, it provides a distinct advantage over traditional, monolithic architectures. For example, a company using a microservice from Google for speech to text in a virtual assistant would find it easy to switch from one microservice to another if Microsoft came up with a better solution. AI architectures should adopt these best practices, allowing solutions to be less complicated, more flexible, and more easily integrated with other parts of a business.

Many of the nearly 2,000 AI-related technology companies today provide microservices. If a company still uses a monolithic architecture, however, and wants to change over, there are software packages available that create a microservices layer around the legacy applications. These microservices layers can integrate directly into the legacy application or can integrate using RPA software as an intermediary. This approach allows the enterprise architecture to be fully microservices-oriented to support AI initiatives within the company without having to replace every legacy application. Eventually, however, legacy applications will need to be modernized.

Sidebar: The Platform That Drives Uber: Michelangelo

Michelangelo, Uber's internal machine learning platform, "enables internal teams to seamlessly build, deploy, and operate machine learning solutions at Uber's scale. It is designed to cover the end-to-end ML workflow: manage data, train, evaluate, and deploy models, make predictions, and monitor predictions. The system also supports traditional ML models, time series forecasting, and deep learning."[1] Uber developed Michelangelo in 2015, and it has been serving the company since 2017. Michelangelo was built because previous AI efforts were inadequate to the task of helping a business of Uber's scale and size. Separate engineering teams had been creating limited one-off systems based on mostly open-source software. There was no standard place to store data, no standard place to store the results of training experiments, no way to compare them to one another, and no ability to train models larger than what would fit on a desktop machine. Most important, there was no standard path to deploy a model into production. Michelangelo changed all that, and it is currently being used across the company in a variety of use cases.

AI Platform Architecture

The major functional components of AI platform architecture should align with the AI lifecycle. These components can be broken up into these four high-level areas: data minder, model maker, inference activator, and performance manager.

Data Minder

Next to a business's staff, data is its most valuable asset. The data minder component ingests data, curates it, and makes it accessible via self-service from across a company, making it available for AI modeling consumption. AI scientists may revise code while creating an AI model, but they make changes only to a curated copy of the data, leaving the raw data untouched. AI scientists should not have to provision data every time they need to create a model; it is a lot quicker and more efficient to aggregate and create an accessible data set and use it multiple times, enabling anyone who has access to share it. This means not having to set up data sets more than once, and it allows for consistency in using data. Also, the data minder component enables teams to work together more efficiently as well as eliminate repetitions of some costly activities, such as copying and extracting data and managing data quality.

A platform should allow for business-specific language for feature selection and transformation. This language enables the use of nomenclature that is familiar to the users of the data rather than presenting them with confusing technical terminology that they do not recognize. The vocabulary is generally implemented through a data catalog and a data lineage tool using business terminology. A platform should allow for the addition of new structured and unstructured data types without the time lag and the cost of expensive data standardization or hardware and software provisioning. It should accurately recognize who has access rights to it. Additionally, when a team has already worked on feature-engineering a dataset for modeling purposes, that set should be added to a feature marketplace; it is a curated collection of engineered features that can be reused across different models.

Model Maker

The model maker component of platform architecture enables the creation of new AI models and faster experimentation while avoiding duplication of effort, automating low-value tasks, and improving the reproducibility and reusability of all work. The environments enable teams to collaboratively maintain, clone, reuse, and extend AI models and support version control. It contains many technology components and supports the different types of algorithms that the AI scientists are using to develop AI models. It must be able to grow and evolve as a business does, giving an organization a structured yet flexible way to create AI-driven solutions today and over the long term.

This component tracks each model easily and efficiently, answering questions such as: Which dataset was used? What algorithms? What hyperparameters? What results were obtained? It has the ability to reproduce identical results when the same model is rerun with the same data. It also allows AI scientists to go back and iterate from an experiment at some point in the past with new data, as well as understand why a particular model worked best.

The model maker enables each function or business within an organization to work in parallel, leveraging the tools and plug-ins they need. It allows for standardization of workflows and tools across teams via an end-to-end system that enables users to build and operate machine learning applications quickly. Standardization enables them to share data, which is especially important since they are likely to be running different models on that data, each oriented to a different use case.

To enable a business to evaluate its AI models, the platform has the ability to track each experiment. Ideally, it autogenerates optimal features, auto-selects optimal models, and auto-optimizes the hyperparameters. It produces outputs, sometimes in a visual form, for a broad range of users, from executives to power users to AI scientists.

Inference Activator

The inference activator component is where the platform supports the propagation of a specific version of a model into production. Its purpose is to get from model development to model deployment as rapidly and efficiently as possible. Model deployment includes the ability to deploy models independently or with dependent code and data. This component enables parallel testing or A/B testing of different models. It does this automatically in all the stages of the model's lifecycle: sandbox, model validation, systems integration test (SIT), user acceptance test (UAT), staging, and production. Each time a model passes inspection in one of these environments, a platform pushes it to the next environment with one click.

A multitude of things may prevent or delay the activation of models, that is, putting models to use in production. One is the lack of automation of the deployment process using the appropriate AI DevOps. Others include managing the model risk, getting user buy-in, and trading off between performance and risk. Even if an enterprise has increased the number of experiments and done many proofs of concept, there is zero value realized if it does not deploy them into production. Zero times anything is still zero. It is critical for end users to access the model output, since these insights are only useful when end users have them, and sometimes only within a certain time frame.

The inference activator has a modular design, so the best intelligent products available for a company's needs can be plugged in or swapped out. Some of these are niche products that have one specific use, such as personalization engines that customize a company's website for each user. Others are plug-and-play, such as a customizable speech-to-text or a sentiment-analysis model that the team can reuse in various applications. A flexible, evolutionary architecture allows AI scientists to combine its multiple functions into a successful system that can handle a company's evolving circumstances. It allows for the flexibility to configure newly available cognitive capabilities and replace or improve existing ones, whether they are models built by the company's team or commercially available components that can be leveraged within the platform. These components might initially be a significant portion of the available models.

A platform should not be overly tied to existing internal IT systems; this way, costly changes to and integration with legacy systems can be minimized and implementation costs of the new cognitive platform contained. The loose coupling can be managed through an API layer. Maintaining this flexibility in the architecture enables the enhancement of the cognitive platform over time. It should also help avoid redundant published APIs: those application program interfaces that tell a business and the world how to talk to that business's AI system.

Some platforms offer integrated dashboard functionality or provide connections to visualization tools, which goes beyond just providing an API for model results. This makes model outputs much more accessible to consume or integrate. However, having users log in to a new system can disrupt their workflow. That is why some companies choose to use APIs to integrate the models into their BI systems or business applications.

Performance Manager

The performance manager component of the platform is where the quality of the model is monitored over time. AI model outputs, such as predictions for events, are captured, as are actual events. Then the two are compared. The differences between the model outputs and the actuals should be within the same performance threshold that was used when the model was approved to go into production.

The performance manager monitors feature drift, so the statistics of the model inputs are not worse than when the model was trained, especially when the actuals take time to gather. Feature drift is acceptable within some thresholds, but if the drift threshold or the performance threshold is breached, the model needs to be retrained. Platforms may also need to have separate workflows for different levels of business risk. The performance manager also supports the auditing of the predictions or recommendations that the AI model is making. These requirements are much more stringent in a highly regulated environment, such as finance or insurance. If regulators want to audit models and see all the decisions made in production over the past three months, a platform needs to support that request.

It is also essential to have systems in place that enable a business to recognize, highlight, and contain rogue output in a model. A rogue output is one that should not be acted on. This component of the platform should have guardrails, sometimes also called *circuit-breakers* or a *fail-safe mechanism* so the system knows what to do in such a circumstance. Guardrails alert users and invoke a fail-safe mechanism, automatically preventing a system from taking specified actions when a model exceeds predefined limits (that is, when results look odd or wrong). It is crucial to have the ability to turn off models and to maintain an understanding of the downstream impact of such actions. The fail-safe mechanism should also involve manual overrides, failover models, or a combination of these. It should include procedures and escalations for handling errors, and any communication needed, internally and externally.

Sidebar: Choosing AI Frameworks

An AI framework is the collection of software and pre-packaged algorithms that work as part of a platform with which you can build trained AI models. Choosing the right AI framework or frameworks is an essential part of creating a suitable AI platform, and by extension, a successful AI enterprise. Each framework has strengths and weaknesses and may change over time. For example, it may be that the toolkit originally chosen is now outdated as newer frameworks emerged. In that case, a company will want to be able to either change the framework altogether or add one or more to supplement the one already running. Alternatively, a framework that has been in use might no longer be supported and is being phased out. For these reasons, a framework should be flexible enough to accommodate multiple current and upcoming AI tools.

A framework should also support the programming languages a business prefers – at the minimum, Python and R, which are currently the most prevalent in AI environments as well as reasonably easy to use. It is a good idea to consider leveraging mainstream toolkits rather than niche products. It is also useful to leverage existing APIs where possible rather than creating new models that duplicate existing ones. Existing APIs have already been tested in the field, and that means they tend to be more reliable. Current popular frameworks include Facebook's PyTorch, Google's TensorFlow, Microsoft's CNTK, Amazon's MXNET, and SciKit-learn.[2]

In Chapter 12, we delve deeper into a reference AI architecture and describe some of the lower-level components and technical patterns. However, let us look next at one of the more critical considerations when building AI into an organization – managing model risk.

Notes

1. Uber Blog (September 5, 2017). Meet Michelangelo: Uber's Machine Learning Platform. https://eng.uber.com/michelangelo/ (accessed September 27, 2019).
2. DZone (January 10, 2018). 10 Best Frameworks and Libraries for AI. https://dzone.com/articles/progressive-tools10-best-frameworks-and-libraries (accessed September 27, 2019).

Chapter 10
Managing Model Risk

> *Artificial intelligence is just a new tool, one that can be used for good and for bad purposes and one that comes with new dangers and downsides as well. We know already that although machine learning has huge potential, data sets with ingrained biases will produce biased results – garbage in, garbage out.*
> *Sarah Jeong, journalist specializing in information technology law*

People may not notice AI in their day-to-day lives, but it is there. As we saw in Part II of this book, machine-learning-based programs now review many applications for mortgages. AI algorithms sort through resumes to find a small pool of appropriate candidates before job interviews are scheduled. AI systems curate content for every individual on Facebook. And phone calls to the customer-service departments of cable providers, utility companies, and banks, among other institutions, are answered by voice recognition systems based on AI.

This "invisible" AI, however, can make itself visible in some unintended and occasionally upsetting ways. Retail giant Target uses AI to understand what shoppers are buying and what to recommend to them, but the tactic backfired when Target sent coupons to a man's teenage daughter that featured nursery furniture and maternity clothing.[1] Sometime after storming into a Target store outside of Minneapolis and castigating the manager for encouraging her to get pregnant, the girl's angry father discovered that Target knew something he did not. "It turns out there's been some activities in my house I haven't been completely aware of," he revealed in an interview. "She's due in August." It was not that Target's prediction about the customer was inaccurate; the problems arose when the store automatically followed up on it.

Then there was Tay, the chatbot released by Microsoft in March 2016. Tay was supposed to have the personality of a normal 19-year-old, but only hours after its release, it turned into a sex-crazed, racist monster.[2] It seems that Tay's users had exploited Tay's learning model and taught it how to say deeply offensive things. Tay was said to have "gone rogue" after it was released to the public. This certainly provided a teachable moment, but it was an alarming one.

Other serious issues were raised when PredPol, a company formed to create and market predictive policing software designed to predict criminal activities in neighborhoods, discovered that its algorithms predicted crime in communities that had a high proportion of racial minorities, whether or not they were, in fact, high-crime areas.[3] Unlike the preceding Target example, these predictions were not sound. In another case of justice gone wrong, news organization ProPublica found problems with a risk-assessment tool that is widely used in the criminal justice system. The machine is designed to predict recidivism (relapse into criminal behavior) in the prison population. Risk estimates incorrectly designated African-American defendants as more likely to commit future crimes than Caucasian defendants.

There is growing concern that AI algorithms introduce bias and discrimination against protected groups as well as the "potential of encoding discrimination in automated decisions"[4] from bias in historical data, such as against minorities in policing, or against women in jobs and mortgage decisions. The lesson here is that artificial intelligence needs careful governance to mitigate model risk, something that is becoming more important as more companies use AI more frequently.

When Algorithms Go Wrong

In 2018, Amazon scrapped some of its AI recruiting software because it demonstrated a bias against women. As reported by Reuters, Amazon's own machine learning specialists realized that their algorithm's training data had been culled from patterns in resumes submitted over 10 years.[5] Unfortunately, during this period, males had dominated the software industry, so most of the resumes used for training came from men.[6] This example demonstrates two essential points: it is critical to choose appropriate training data, and humans can easily make mistakes in doing so. People can also introduce problems in the case of AI algorithms themselves, despite computer scientists' attempts to do the best job possible given the tools available to them. This may result in AI algorithms that suffer from bias or other issues. It is worth examining these situations and how they can be addressed.

Software in and of itself is neither good nor bad. It can become so either due to how someone uses it, for example, when malware is created with malicious intent, or when its design creates unintended negative consequences. Unintended consequences were less of a problem in the past, given that requirements were written and then hand-coded, reviewed, and tested. Even then, there were some – some in the coding process and some due to gaps in design. AI learns from existing data without that level of human intervention, and depending on a variety of circumstances, this can lead to further problems. For example, the data used for training might reflect human bias in an organization. Likewise, the use of only partial data for training might create a bias that could perpetually codify problems into AI-driven business practices. These will be in addition to the previous challenges with software.

Computers interpret algorithms literally. They do what they are asked to do or, in the case of machine learning, what they have been trained to do, without any application of common sense. Computers do not behave like human beings. They have no judgment outside of their lines of code or learnings from data. There are numerous examples of computers trading off long-term benefits for short-term gains. Websites that use machine learning to improve ad click rates, for instance, without considering customer satisfaction can leave their users drowning in useless and often offensive click-bait articles because their AI model was trained on the wrong objective.

Algorithms are also greedy – greedy being a technical term in this case. These are algorithms programmed to make the best possible choice at each step of some process without looking into the future to predict where all these paths lead. Sometimes this provides the best answer; sometimes it does not. An analogy is a GPS navigation system that optimizes for time above other considerations. Such an algorithm, designed to optimize for even the minutest advantage, would direct a driver to circuitous routes that use more gas or are even more dangerous, as long as it provided even a tone-second lead over more straightforward directions. Even worse, some of these turns may take longer than estimated by the algorithm, taking away any advantage.

These issues are compounded by the fact that many AI models built on machine learning are not easily interpretable. These black boxes take input and provide output or predictions with little or no indication of how or why those predictions were made. Since the computer is "creating" the logic based on data from which it is learning, it sometimes can be challenging to predict how it will behave in any given situation, or why it behaves in a certain way. Even if the explanation can be understood mathematically, few business users have the background to grasp it intuitively. Moreover, there are only limited ways to query an AI algorithm about why it made the decisions it did.

The more that people see AI as a black box, the more they become concerned by its nature. Already we are seeing the public, as well as regulators and the media, worry that AI is a genie that cannot be controlled. These fears will only be exacerbated over time as AI becomes more ubiquitous. As AI gets deeper into financial markets, medicine, law, manufacturing, and everywhere else, people need to know how, what, and why AI algorithms do what they do and how to mitigate the risks of using models. Otherwise, fears will stall applications of AI within businesses.

Mitigating Model Risk

Managing AI model risk is not just a nice idea, but also a business imperative. It is used not just to abide by the laws, but also to do the right thing. As company leaders, employees, citizens, and stakeholders in business and social outcomes, we must think about and act on model risk governance for AI models at every step of their development and use. To mitigate the risks of AI models, companies must put *model governance* in place throughout the modeling lifecycle: before the model is developed, while the model is being developed, and after the model is built.

Before Modeling

Before modeling, focus should be on the data. Teams must ensure they have the right data from which to build their models. Good data requires data quality assessments, such as looking for missing values, systematic errors, and delays in data availability. Governance of data for modeling should include an understanding of data quality; data lineage (knowing what data is being used and where it came from); data privacy (making sure that the model being considered is in line with data privacy based on policy); data propagation processes (to prevent, for example, test data leaking into a production environment); and tracking usage of restricted attributes, such as gender or race. If teams use gender or race in a model, the model needs to go through additional validation or governance steps, such as fairness testing, which we will discuss later in this chapter. Teams also must map regulations to entities and attributes and flag restrictions as appropriate.

One important aspect of good data is that it needs to have breadth and depth. In customer data, for example, data breadth involves looking at a large number of customers, whereas data depth involves focusing on more data about each customer. Broader and more in-depth datasets enable models to handle situations better and more predictably and help reduce bias; in fact, it was this lack of breadth in data that Amazon had to deal with in its recruiting software. One way to think about this is that data depth and breadth add a bit more "common sense" to the learning algorithm.

Fairness also needs to be an objective in algorithmic decision-making. Although using actual data and algorithms can help eliminate existing biases, it can also introduce new biases or codify existing ones from historical data. To address this, recent AI research has tried to define what "fair" is in predictive modeling and how to address fairness. The two primary approaches that companies have begun to use in their business are *group fairness*,[7] which requires that the outcome of a prediction should be similar for all groups, and *individual fairness*,[8] which requires that similar individuals be treated similarly irrespective of group membership.

One way to address fairness before modeling is to assess the data to look for bias, and preprocess the data to remove bias before creating a model from it. There are a few approaches to doing this. One, called *suppression*, removes the protected attribute, such as gender, race, religion, or disability, from the training data. Simply removing these variables, however, turns out to be ineffective, doing nothing to address the bias and even possibly concealing it. That is because this "fairness through unawareness," as it has been called, ignores *redundant encodings* – ways of inferring a protected attribute from unprotected features. For example, if we are building a model to approve loan applications, removing the gender of the applicant as input is often not sufficient, because, based on historical data, if we know the other inputs and the final decision, we can often predict the gender quite accurately. This predictability means that bias has been encoded in the other variables. For instance, zip codes or names can be used to deduce race, gender, or nationality.

To address this, we first look for all the other features and attributes that are highly correlated to the protected attribute (and which could be used to predict the protected characteristic). Then we preprocess the data to remove the bias using approaches that include *data transformation*, in which the input (historical) data is transformed using an optimization algorithm to remove discrimination, and *data reweighting*, in which weights are assigned to the input dataset to make the training data discrimination-free and ensure fairness during model training. In each of these preprocessing options, the decreased discrimination comes at the cost of accuracy and must be optimized as a trade-off. Although I am specifically using the example of fairness here, each of these approaches applies to any type of bias built into the data – whether the data is about people, or about machines from sensors, or trading data about companies. Bias must be identified and addressed in the data before modeling.

During Modeling

While modeling, it is necessary to validate the predictive performance of the model. Performance thresholds such as accuracy, false positives, and false negatives should be defined early and compared to benchmarks. There is often a trade-off between these, so selecting which to optimize for is essential. Recall that in the cancer detection example from Chapter 8, a false negative may have potentially life-threatening consequences, whereas a false positive will not be dire, although it will almost definitely provide the patient with a bad experience.

It is also essential to use an appropriate or diverse set of objectives to train an AI model to avoid the trap of narrow, literal goals. Consider the click-bait example. It would be preferable to create a model designed for user satisfaction or actual purchases rather than just ad clicks. However, to do that, a model would have to be trained based on objectives of user satisfaction, measured, for example, by the amount of time spent on the clicked-through page, or actual purchases rather than just ad clicks. In other words, the goal (i.e., the target of the model) would have to change, at least in part.

There are many situations in which a team may need to incorporate such multiple goals into an AI system. For example, a business might want to increase online sales and have low return rates.

That means it will likely consider a learning target such as number of sales minus the number of returns or the equivalent dollar value. If in the click-bait example we did want to also increase clicks because we want the user to visit multiple pages, we could include time spent and click count as objectives by defining a new objective target, which is the multiplication of click count and time spent on the resulting page. This will help discount scenarios where the user always goes to one page and reads it for a long time and does not go elsewhere, for example. Models built based on multiple goals, using the different dimensions of data variety, decrease the likelihood that these models will single-mindedly try to solve for only one thing. However, in the process, data that is not appropriate must be avoided. What is inappropriate depends on the use case. For example, for someone performing a medical diagnosis, the gender of the patient and the patient's age may be extremely relevant to improving the model's predictive diagnostic power. Including the same gender and age-related data in deciding on a mortgage or loan application, on the other hand, may be unethical or even illegal.

For addressing bias during the modeling process, once the bias from the data has been removed before modeling, one common approach is to postprocess the output – that is, to take the output of the machine learning model and process it further using another algorithm to reduce bias, thereby improving fairness for members of protected groups. Another approach is to add a constraint, called a *bias regularization* term to the model training optimization objective. Usually the objective just includes the error, and in this case, it would be the sum of the error and bias that should be minimized together. Another variation of this that has been recently developed, called *equalized odds*,[9] constrains the algorithms so that neither false positives nor false negatives disproportionately affect any protected subgroup. There are other alternatives and variations that the AI scientists should explore depending on the business need for low bias.

It is critical to check the model-making process and data usage to ensure *reproducibility*. Machine learning attempts to find the function with the least error. If the modeling process does not find the global minimum but just the local minimum, then each training iteration could discover a different minimum and, hence, end up with a different prediction. Imagine a hiker who is looking for the lowest point in her area. She reaches what seems to be this low point, known as a minimum. If she moves a small distance in any direction, she's heading uphill again. However, this low point may only be a local minimum, not a global minimum: there may be other valleys (local minima) in other locations that are lower than the one she is in. In the algorithmic equivalent of this situation, the training process may have found a local minimum of the error function but not its global minimum. Rerunning the training process can change the resulting algorithm significantly if it finds one local minimum and then, upon retraining, finds a different one, hence giving different results.

The modeling process should also include code reviews to both ensure best practices and confirm that models or parts of models can be reused across teams.

After Modeling

After a model is built and validated, it is critical to perform a variety of model risk tests, including compliance testing, fairness testing, sensitivity testing, and boundary condition testing. A model also must be checked for interpretability of its results.

Fairness testing looks at whether underprivileged or protected groups based on such things as gender or race are being treated fairly. This is the same as bias testing for any other type of data. This ensures that the final model is not biased. It tests for how the model behaves with different inputs – for example, if the same data is utilized when determining the results of a mortgage application but the gender is changed, does the output or decision change? It also tests for aggregate results – for example, do as many males as females get approved or rejected on the new dataset?

Most of these techniques can be evaluated based on various algorithmic fairness metrics such as equalized odds, statistical parity, disparate impact, predictive parity, equal opportunity, average odds, or predictive equality. However, both fairness metrics and accuracy metrics need to be defined during modeling, as some may work better together than others.

Sensitivity analysis validates that different combinations of input fields are behaving appropriately. Sensitivity analysis and its relative, *boundary condition testing*, involve providing a variety of perturbed inputs (where the inputs are changed slightly) and examining the outputs to discover which input factors are most important in determining specific outcomes and whether, under some input conditions, the model can "go rogue" by providing potentially invalid outputs. Sensitivity analysis can also illustrate how a model's outputs are dependent on its inputs in different regions of the data input space.

There are a variety of ways to do this. First, teams need to understand the distribution of the values of features. Say the two features with which a business is dealing are age and height. The team begins by summarizing the data, determining the minimum value (e.g. the lowest age and height), and the maximum value (e.g. the highest age and height). Then it determines the average, or 50th percentile value of each. From there, the team calculates other values that might be useful, typically, the 25th percentile of each feature (the age or height at which a quarter of the data points are lower) and the 75th percentile.

Now the team establishes a reference point by running the model using the average values of its features and then changes the value of one feature at a time (say, age), while keeping the other one (height) at its average value. The team can then keep changing age to different values, such as its minimum, its 25th percentile, its 75th percentile, and its maximum, and observe how a prediction varies. It can also alter the values of two features at a time rather than one, or in the case of more variables, try all combinations at the same intervals.

Say an AI scientist is using the features of age and height to predict body weight. To see if the model is giving reasonable responses under different inputs, she runs the model on the average height and age, say, 5'10" and 45. After fitting the model to the data, her model predicts that, a 5'10" and 45-year-old individual has a body weight of 150 pounds. To determine the sensitivity of the model, she might then run it using the same value for height but changing the value of age to 44, then 46. This tells her how sensitive the predicted weight is to a small change in age. Now she can use the 25th, 75th, or other percentiles. If the output does not change much, even under substantial modifications to the input, she has learned that age is not essential in predicting weight. If it does change, then age is an important feature in the model. Moreover, it will indicate the particular point at which the model starts to become more sensitive.

Boundary condition testing involves using extreme values of input as test cases. This may mean changing the value of the feature age more dramatically, for example, to the 95th percentile or even 99.9th percentile. An unexpected output, such as a body weight of 3,000 pounds in this scenario, would indicate the model has a problem. The team needs to find a better model that can be relied upon to perform well. They must make sure that no combinations of reasonable but extreme inputs give an unreasonable output.

It is useful to use *model interpretability* to develop an intuitive understanding of a model. Model interpretation is the practice of developing a post hoc explanation of the results produced by the AI model. *Interpretability* is a somewhat vague term that means the ability to view what a computer is thinking in a way that is understandable to a human. Since there is no precise characterization or common understanding of interpretability, people have proposed definitions such as: a low-level mechanistic understanding of models, applied to the features, parameters, models, or training algorithms; or, a model that uses cause-and-effect to justify its decisions. This way of viewing model interpretability is sometimes called *justifiable AI*.

It is more useful to think of interpretability as something that enables users to combine the model's information with their own knowledge. This allows them to start to understand the model outputs, to develop better intuition, and from there to make changes to the model or refine the explanations that may be needed to implement the recommended outputs (for example, why a particular trading strategy should be followed). This can be very useful even though it does not shed light on a model's exact inner workings. In fact, in practice, it is usually more helpful than understanding the precise mechanism of the algorithm. For example, if a model is predicting churn for a given customer, a diagnostic model may be able to provide intuition to a user by pointing to similar customers who have already left. In this case, *similar* means that the information on this customer in terms of usage, calls to a call center, and so forth, resembles that of other customers who have churned. Or the diagnostic model may be able to say that churn is predicted because the customer's calls to customer care has increased significantly. Building model intuition is like building driving intuition – a driver does not need to know the detailed mechanics of how the car works but does need to understand how it behaves in response to the road and in response to her inputs.

Although interpretability continues to be an active area of academic research, a few approaches are being productively implemented. Businesses today are utilizing these techniques to address a variety of issues. This enables teams to successfully govern models so they can continue to deploy them into production. One such method involves using *surrogate local models*,[10] which are models that closely mirror the original model but are more interpretable, although they may be less accurate. An analogy to understand surrogate models is to think of how to approximate a complex polynomial function using linear surrogate functions in different parts of the input data space (i.e. the x-axis). Figure 10.1 shows a polynomial being approximated by five surrogate models. The first one is only valid for input data (x-axis) from 0 to 1, and the second one is only valid for input data from 1 to 2, and so on. You cannot assume the first surrogate function (straight line between 0 and 1) will work well for all data inputs (e.g. for $x = 3$). Also, each surrogate model is less accurate than the polynomial. However, the linear functions are more intuitive and interpretable.

Similarly, machine learning models that work for all the data are complicated multidimensional functions that are not necessarily easily interpretable. A way to build intuition about them is to create multiple simpler models, using different subsets of the data that are more interpretable.

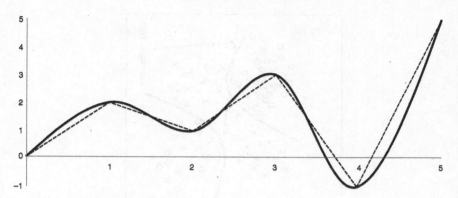

Figure 10.1 Approximating a polynomial function using simpler linear functions in different parts of the x-axis.

Figure 10.2 gives a schematic of how the surrogate model approach works in practice. This example shows a potential classification problem, say with X representing customers who may face financial distress in the next 12 months and O representing customers who will not. The AI model is trained on a sophisticated black box algorithm, and the X-and-O classification is the output. The classification model is shown in the bold lines in box 1 that separates out the Xs from the Os.

To try to understand how the decision was made for a specific customer, the input and output values of her case are selected for interpretation (box 1). Now, a surrogate model is created around this point by identifying nearby points (box 2). These are then perturbed, or changed, by a small amount, and the perturbed data is used as input into the original black box model to get new outputs. These inputs and their corresponding outputs are used to train a more straightforward, usually linear, model (box 3) in which the perturbed inputs are weighted by their proximity to the instance being explained (the original record selected). This simpler linear model is then used to identify significant features and their ranked contributions to the classification. These parameters may serve as an explanation for the result in question.

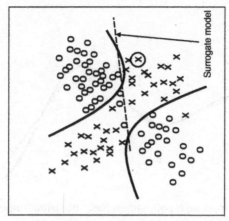

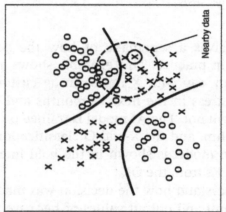

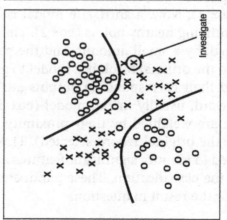

Figure 10.2   An example of how surrogate models can help with interpretability.

Notice that this new model works as a local boundary for the two types of classification of X and O *near* the identified data point under investigation. It is important to remember that this model does not work globally. For example, if you move toward the center or left in box 3, the surrogate model's dividing line and the original model's dividing line diverge significantly. In the surrogate model (dashed line in box 3), in the vicinity of the original data point, anything above the line is O and below the line is X. When you move to the left, this surrogate model will now give incorrect predictions, because it will still predict anything above the line as O. However, we can see that the real data shows X in this area. The surrogate model will only work in the vicinity of the initially identified data point. If you need interpretation for a different region of the graph, the process will have to be repeated for a different initial data point.

For each of these options regarding interpretability, the AI team should also consider the ease of use of the output through interactive visualizations. This will go a long way in getting end users to understand the models in a heuristic sense. At the same time, AI teams must keep in mind that they should avoid putting unnecessary limits on their AI models. They want the algorithm to make accurate predictions, often beyond any human's capacity; however, they do not want to take in any unwarranted bias or risk or disqualifying opacity. They are treading a fine line here, and a certain opacity may be in order. Trading off too much predictive power to increase interpretability may do a disservice to the use case by limiting the full potential of AI applications.

Sidebar: Model Interpretability in Deep Learning

Deep learning networks do very well at assigning classifications to samples, such as cat, not-cat. However, deep learning gives very little insight into the algorithm's inner workings. While deep learning interpretability remains an

(continued)

(continued)

active area of research, recent studies have been promising. As discussed in Chapter 2, the multiple layers of these networks take inputs and pass them from layer to layer, providing higher and higher-level descriptions the further they are from the input layer. Sometimes, when you get closer to the output layer, these higher-level descriptions may track to things that resemble the way a person would describe them. They might, for example, have nodes or groups of nodes that seem to indicate features like fur, tails, or ears. Testing has shown that for specific tasks, such as classifying images, looking at the outputs from each layer helps provide users with a better understanding of how the model is functioning.[11]

Once the AI model gives its output or makes its prediction, the team must think about how best to use that output. Reconsider the Target example discussed earlier. The Target team looked at prior customer purchases to predict other products they should recommend. However, they did not have filters that prevented them from getting negative press. It may not always be obvious what kind of filters to use, and it requires thinking through different scenarios to determine when to trigger what types of actions (or not) from a model output.

In an attempt to provide transparency to individuals affected by algorithmic decisions, the General Data Protection Regulation (GDPR) (see Chapter 7) gives people the right to an explanation when confronted with an algorithmic decision such as why they were turned down for bank loans. As one might imagine, this can be problematic due to the complex nature of the input and the opaqueness of AI algorithms. Often, there may be no ready explanation for an algorithm's decision. GDPR's Article 22 states "Automated individual decision-making, including profiling, is made contestable."[12] This means that a company must be able to interpret model outputs and prove that its algorithm is not making false claims against a customer.

This legal right to an explanation requires transparency, and some companies in Europe and elsewhere have revealed some portions of their source code as a result. However, this provides only a small glimpse of what the code is capable of in terms of bias. Moreover, as mentioned before for machine learning, the bias may be in the training data and not in the code at all. Therefore, model interpretability is a must-have requirement for businesses that need to comply with GDPR. These regulations apply to all multinational companies doing business anywhere in Europe. Also, since they require a retooling of privacy standards, companies should consider applying them to the rest of their customers around the world.

Model Risk Office

When defining and putting in place a model governance approach, consider the full AI model lifecycle, as discussed in Chapter 8. For a healthy, independent, repeatable model governance process, a business must set up a model governance office or team. The consequences of not doing this can be significant. Consider the example in which a regional bank created a valuable AI model. Executives immediately wanted to put it into production, which would have allowed them to do more cross-selling and up-selling. However, the CEO was concerned: "I don't know what it is going to do. What if it goes rogue? There's too much risk to our reputation."

The problem was that the bank had neglected to create a governance process for AI algorithms. It took two and a half months to develop their AI model; it took nine months to set up a governance system and use it to approve the model. Had they begun to set up this governance system before working on their first algorithm, it might have taken as little as two days to take the new model through the governance steps. It is more efficient and cost effective to set up governance processes early, making sure they apply to any AI algorithm that is developed.

Not every use case needs to go through the same level of rigor in terms of governance. For a bank, one use case may be to personalize the content on the website, and another may be to forecast cash flow and interest rates. Getting the first one wrong may have a lower consequence, whereas getting the second one wrong could cause liquidity problems for the bank. The team should decide which governance steps the model should go through by taking into consideration the likelihood of issues and the level of impact or consequence that results from those problems. If each model is put through the highest level of rigor, it will unnecessarily slow down low-risk models from going into production and potentially take away focus from higher-risk ones. Different factors that are specific to a given business may apply, and companies should develop a *model risk framework* to decide the level of risk for different types of models within their business context. This framework should consider the level of consequence for the model types and uses, and prescribe what governance needs to be in place for different categories of models during each step in the lifecycle for each level of risk (high, medium, or low). The framework should define which steps are applicable for each model category, and which steps to drop to gain speed and reduce the cost for moderate and low-risk models. Businesses then need to decide, for each model they develop, what risk category it falls in within the framework, and hence what governance steps that model needs to go through.

For fairness testing, there should be a set of standardized validations. Teams must check to see if the model is discriminating against protected groups of individuals (such as those who are members of the categories mentioned earlier of gender or race). It is also critical that the model be compliant with regulatory requirements.

Before hooking up the model to a production process and triggering its use, the team should ensure that the model is working well with upstream (usually data) and downstream (usually usage) systems. The team should also validate that there is a failsafe mechanism in place so that if something goes wrong with the model in production, there is something to fall back on, either another model or some fallback course of action. If a model is found to not work well, the team might revert to a prior model that is known to work, or a simpler rule-based alternative that will work temporarily. Depending on how the model is used, another alternative may be to increase the frequency of manual interventions in a decision – for example, if the model output is stored before it is used (batch models), then these outputs can be reviewed for bad results and overridden manually, until the model is fixed. If the failsafe alternative is another AI model, that model also has to be compliant with the governance steps. It is a good idea to consider these and other failsafe mechanisms from the outset.

In addition, other situations may need to be addressed, such as a model starting to degrade in production or no longer working as expected or desired, for example, because the statistics of the input data have changed from the training dataset. Ongoing monitoring of the model performance in production must be in place to find these situations. This includes capturing both what the model is predicting and the actuals as they become available. That way, someone in the company can detect and appropriately handle any degrading of model performance quickly. Sometimes, this can trigger the failsafe mentioned earlier, and sometimes it requires the model to be retrained.

All of these aspects of risk mitigation are important to consider before implementing AI models at scale. Next, we will look at how an organization sustains and grows AI capability by considering the best organizational structure for its implementation.

Notes

1. *Forbes* (February 16, 2012). How Target Figured Out a Teen Girl Was Pregnant Before Her Father Did. https://www.forbes.com/sites/kashmirhill/2012/02/16/how-target-figured-out-a-teen-girl-was-pregnant-before-her-father-did/#20e4736b6668 (accessed September 29, 2019).
2. *Wired* (February 13 2017). How to Keep Your AI from Turning Into a Racist Monster. https://www.wired.com/2017/02/keep-ai-turning-racist-monster/ (accessed September 29, 2019).

3. *New Scientist* (October 4 2017). Biased Policing Is Made Worse by Errors in Pre-crime Algorithms. https://www.newscientist.com/article/mg23631464-300-biased-policing-is-made-worse-by-errors-in-pre-crime-algorithms/ (accessed September 29, 2019).

4. Executive Office of the President (May 2014). Big Data: Seizing Opportunities and Preserving Values. https://obamawhitehouse.archives.gov/sites/default/files/docs/20150204_Big_Data_Seizing_Opportunities_Preserving_Values_Memo.pdf (accessed September 29, 2019).

5. Reuters (October 9, 2018). Amazon Scraps Secret AI Recruiting Tool That Showed Bias against Women. https://www.reuters.com/article/us-amazon-com-jobs-automation-insight/amazon-scraps-secret-ai-recruiting-tool-that-showed-bias-against-women-idUSKCN1MK08G (accessed September 29, 2019).

6. Ibid.

7. Michael Feldman, Sorelle A Friedler, John Moeller, Carlos Scheidegger, and Suresh Venkatasubramanian (2015). Certifying and Removing Disparate Impact. In *Proceedings of the 21th ACM SIGKDD International Conference on Knowledge Discovery and Data Mining*, Sydney, NSW, Australia, August 10–13, 2015, pp. 259–268. ACM (accessed September 29, 2019).

8. Cynthia, Dwork, Moritz Hardt, Toniann Pitassi, Omer Reingold, and Richard Zemel (2012). Fairness through Awareness. In *Innovations in Theoretical Computer Science 2012*, Cambridge, MA, January 8–10, 2012, pp. 214–226. ACM (accessed September 29, 2019).

9. M. Hardt, E. Price, and S. Nathan (2016). Equality of Opportunity in Supervised Learning. In *Advances in Neural Information Processing Systems* (accessed September 29, 2019).

10. SIGKDD (2016). "Why Should I Trust You?" Explaining the Predictions of Any Classifier. http://www.kdd.org/kdd2016/papers/files/rfp0573-ribeiroA.pdf (accessed September 29, 2019).

11. Cornell University (November 28, 2013). Visualizing and Understanding Convolutional Networks. https://arxiv.org/pdf/1311.2901.pdf (accessed September 29, 2019).

12. General Data Protection Regulation (May 4, 2016). Automated Individual Decision-making, Including Profiling. https://gdpr-info.eu/art-22-gdpr/ (accessed September 29, 2019).

Chapter 11
Activating Organizational Capability

Like all technologies before it, artificial intelligence will reflect the values of its creators. So inclusivity matters – from who designs it to who sits on the company boards and which ethical perspectives are included.
Kate Crawford, professor at New York University and researcher at Microsoft

Thus far, we have looked at the core processes and tools that a company uses to implement an enterprise AI strategy. However, it takes more than processes and tools to grow and sustain a company-wide AI capability; it requires organized people. Ali Ghodsi of UC Berkeley has said, "Only about 1% of companies are succeeding in AI. The rest are struggling with three problems, the first of which is getting people to work together and collaborate with one another."[1] There are significant barriers to embedding AI within most organizations; some aspects of the company's operating model will have to be reconfigured to successfully implement AI at scale. To prepare the organization for using AI at scale, the company needs to help their management teams, business employees, and others to think in new and different ways. But even before this, the first thing that must be done is to achieve stakeholder alignment around AI.

Aligning Stakeholders

Successfully building an AI capability within an organization requires strong support from the board and the management team, something that is true for building any organizational capability. The trouble is, management may feel as if it is betting (sometimes the company) on a technology that has not yet been proven within the organization. This feeling can be unsettling, and to overcome it, executives should take the time to understand how AI and machine learning work; such executives need to enable better decision-making by building their own intuition for what in AI is real and what is hype, and think about the steps necessary to implement AI solutions across their organizations. The management team can then define a vision for AI. This book can serve as a guide while developing that vision.

Critical to the success of any AI program, however, is the support not only of the executives but of most if not all other stakeholders. Leaders must unite their organizations behind their visionary objectives. Employees must grasp why AI is important to their company and understand both how AI will affect their jobs and how they can contribute to an AI-centric organization. Executives will need to explain that jobs will likely change, but for the better, improved by the addition of new tools based on AI. Business leaders need to emphasize that AI does not replace most jobs; it replaces tasks, often onerous ones, allowing employees to focus on strategy and creativity instead of work that is routine or boring. Negative AI stereotypes must be countered to allay fears of loss of authority and promote a healthy work environment in which both humans and machines work together and thrive.

Leadership should be prepared to face some resistance. Implementing an AI strategy can involve significant change, and that can make people uncomfortable. In this situation, it is best to start with incremental changes or a pilot project within each function or department to get groups comfortable with AI. This means reaching out to personnel who may not regularly interact, such as those in marketing, the supply chain, HR, and IT. Additionally, employees in the company must understand how to participate – how they can identify potential use cases, what steps they need to take to initiate the building of solutions, and how they can leverage AI in their roles. The result of this should be that there is alignment around AI goals and growing interest in contributing to AI solutions across different parts of the business, from front to back office, across departments and brands.

Organizing for Scale

Once there is alignment on the vision and goals of AI, the leadership team needs to determine where in the organization the AI teams sit and how they interact with other groups. There must be a well-understood organizational structure that includes business, IT, and AI teams and that designates associated roles and responsibilities for successfully executing and maintaining AI projects. The various functions need to collaborate across organizational structures comfortably. Without this collaboration, a business cannot move forward. Choosing the best structure can be a delicate balancing act, since there may not be consensus about where AI and analytics capabilities should sit within a business. Moreover, the best structure is specific to each organization – something that works for one company might not necessarily work for another (see Figure 11.1).

One approach is to have a *centralized structure* for the AI and analytics teams, under a single leadership. In this structure, AI scientists report to the chief AI officer or chief analytics officer and are assigned to projects as needed. Centralization encourages collaboration and knowledge sharing among AI peers. It leads to better adherence to the company's AI policies and greater standardization of applications across the business units by ensuring that everyone in the group is working from the same set of AI rules. It can also allow for the fast-tracking of strategic AI priorities because planning and execution are centralized.

However, a centralized structure means the AI scientists are not "in the same room" as business users – they are somewhat distant

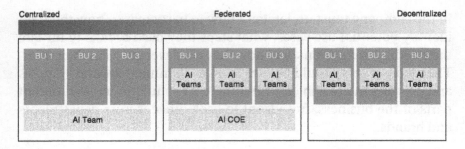

Figure 11.1   Centralized, decentralized, and federated operating models for AI.

from the decision-makers in the business units and from the users operating the business. This may hamper the AI team's ability to successfully execute on use cases because it is harder to get input from business units and get buy-in to use the models once they are developed. A centralized structure also has other drawbacks. Initiatives may slow down if they have to go through a centralized team every time a business unit wants to launch an AI project. Also, if a business unit is assigned a different AI scientist each time it begins a project, the resulting lack of continuity can cause the business unit to feel that the centralized structure does not fully support their needs.

An alternative approach is a *decentralized structure*, in which individual business units build their own AI science and engineering capabilities and hire their own teams. Each business unit sets its agendas and priorities independent of other business units and is responsible for solving AI challenges in its own way. In this model, AI scientists get their instructions from business units directly, enabling the AI team to acquire a deeper understanding of that business's requirements and nuances. It allows the AI team and business team members to have more frequent and open communication. It makes the AI teams feel invested in the process. The closer ongoing interactions often increase business buy-in to the outputs from the AI teams.

Decentralized models can have their drawbacks. For instance, they require every business unit to have managers who have enough experience and facility to both hire the right AI scientists and to guide them in their work. Also, decentralization may lead to duplication of effort, not allowing the organization to leverage what has already been built and resulting in a variety of similar projects being managed across business units. Another disadvantage is that individual business units may only be able to solve the smaller problems, not having the individual budgets or capacity to take on the more substantial challenges across the business.

Consider the example of a company in which the CTO discovered that six different chatbots were being developed among different business units with significant overlapping functionality, all at varying stages of completion. Every one of these units only focused on solving the easier problem of intent recognition in the user questions, avoiding the more complex problem of using AI to find answers to the questions automatically from unstructured text. This redundant effort could have been avoided if group planning and coordination had been addressed at an earlier stage.

A third option is to take a hybrid approach – a *federated structure*, or hub-and-spoke model, that takes advantage of the strengths of both the centralized and decentralized models while minimizing their downsides. This federated model has a central hub – we will call it an AI center of excellence (CoE) – and spokes that sit inside different business units. In a structure such as this, most AI scientists work in the business units but coordinate through the AI CoE, so they get the visibility necessary to potentially cross-leverage knowledge and solutions.

AI Center of Excellence

The AI Center of Excellence takes responsibility for those parts of the AI work that benefit from being centralized (see Figure 11.2). The AI CoE is accountable to both upper management and to the business units, for which it provides guidance and support.

Some examples of the functions that the AI CoE covers in many organizations include the following:

- Education of employees and getting their buy-in to use AI within the organization.
- Setting standards for AI projects across the AI lifecycle and making sure that projects are adhering to these standards.
- Sharing knowledge and building an AI community.
- Defining and coordinating training and helping hire AI talent across the company.
- Owning and growing the AI platform.
- Maintaining awareness and being the central point for the company's participation in the external AI ecosystem, including large technology companies, the AI startup space, and AI academic research trends.

These are covered in more detail in the following subsection.

Standards and Project Governance

An AI CoE is responsible for setting standards for how an AI project is run and for ensuring that relevant standards are followed for

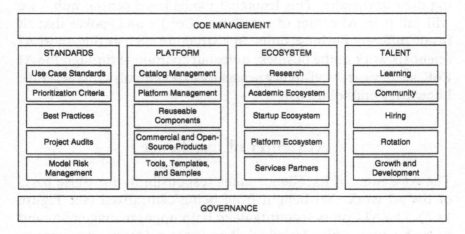

Figure 11.2   Key functions within an AI center of excellence.

each project. Often convened by the chief AI officer, the CoE can be instrumental in building momentum for an AI initiative. It consists of a governance team of leaders including from the organization's business units, and it is designed to ensure that project teams, distributed between the CoE and business units, collaborate and share accountability, regardless of how roles and responsibilities are divided. For an AI CoE to work well, a governance structure must be in place that ensures this accountability and providing the tools each team needs to make good decisions and build robust solutions. A well-planned governance structure fosters collaboration among members, allowing them to share knowledge with their peers.

The AI CoE should be authorized to require projects to participate in project reviews and provide evidence that its standards are being met. In a federated model, in which the business units (BUs) set the priorities for their various AI projects, the AI CoE functions as a facilitator, providing help for such things as managing vendor assignments for AI-related resources. One key aspect of its role is to look at reusable components across BUs and provide advice on the optimal sequencing of projects to get the most benefit from shared components that are built on the platform.

Early on in a company's journey, most AI projects are reviewed by the AI CoE to help make sure that the teams are adequately staffed with the right skills as well as to ensure that all the traps that can derail AI projects are being taken into account. These early projects set the standards for monitoring and managing the performance of the various AI applications that are built. Each BU or group monitors the applications that it owns. In the area of model governance, the CoE sets the criteria and publishes a set of standards that teams must follow, which includes ways to determine if any given project team is in compliance. Teams then have to demonstrate to the center of excellence (or another group within the BU) that they are doing all the mandated work, which might include use case harvesting and prioritization, data quality assessments, AI lifecycle processes, fairness testing, leveraging the platform for modeling and monitoring, and conducting model risk assessments, perhaps by showing test results themselves.

Community, Knowledge, and Training

To successfully implement an AI strategy not only requires excellent AI scientists who understand the businesses of which they are a part; it also requires employees, from CEOs on down, to understand enough about AI and what it has to offer. This dual understanding goes a long way toward making an AI venture successful. AI can delve deeply into the analysis of a process, whether it is how customers are reacting to a new ad campaign or how an organ delivery vehicle can save precious minutes by taking a different route. This capacity requires an environment in which everyone from top management to the people on the shop floor participates in identifying relevant use cases for AI. A lot of the innovation that allows AI to be useful happens inside a process, where employees have the best understanding. The better educated they are about AI and the more comfortable they are with its use, the more their company reaps its benefits.

The AI CoE should encourage feedback loops in which shared learnings throughout the organization are collected and dispersed in order to strengthen the BUs' AI capabilities over time. These loops ensure that everyone, including those in teams within the BUs, has the latest knowledge on what is working, how to address specific challenges, what new tools and products are becoming available, and what are innovative best practices.

The AI CoE should coordinate across BUs to understand and define the training needs for the organization. Training needs to be specific to each audience. They should cover general awareness and how projects work for all the business employees, how to deploy and leverage AI models within applications for the IT teams, and advanced AI topics and business knowledge for the AI teams. General employee training should include how to understand what an AI project looks like; how to identify potential use cases; the processes for initiating implementations for them; and what roles employees should play as projects move from conception to deployment and beyond. The best AI education is a combination of in-person and online self-paced training. However, this plan cannot just be an ad hoc list of point training courses and executive programs. It needs to be approached as a full scale, broad capability building exercise. One recent example is from Amazon, who said they are planning to invest $700 million to train employees.[2] Businesses should convene the executive team for seminars. Ambassadors should go "into the field" to encourage everyone to get on board. This training and preparation will mobilize employees to best leverage AI.

In addition to all this, CoEs can make the company a more attractive place in which to work, which can help the business hire the best AI talent available. Often, AI team members can rotate in and out of the AI CoE when needed, allowing them to move between business units and thus providing them with a variety of new problems to solve. CoEs also help to maintain a stronger community of practice, with AI scientists and engineers distributed across business units and in different locations.

Platform and AI Ecosystem

The AI CoE, together with IT departments, should own, maintain, and enhance the AI platform based on business needs across BUs. It helps coordinate which components from different projects should be built as reusable components so that future projects can leverage these for speed. In some cases, the AI CoE may take on projects before they are needed within the BUs, especially if these involve building complex components that can be shared across BUs. More often, CoEs develop accelerators on the platform that every BU can use to speed up their projects. Examples of these are a framework for model risk management or a tool for automated data-signal discovery.

No company using AI is on its own, but rather exists in an ecosystem of AI activities. The AI CoE should maintain a connection with this ecosystem to ensure it is always making the best decisions about tools, frameworks, methodologies, algorithms, or applications. This ecosystem management should include maintaining relationships with the major technology firms such as Microsoft that are spending billions of dollars on enhancing their AI platforms; with startups that have applications that are relevant to the company; with consulting and services companies that can provide team members and knowhow; and with universities and other research institutions that are developing new algorithms and models to solve new types of problems. An added advantage to maintaining university partnerships is that it is another way to recruit top talent, enabling companies to identify and develop relationships with promising AI scientists. It is also a good idea to monitor sites such as Kaggle, which promote competitions among both amateur and professional AI teams. Kaggle provides teams with identical data and challenges them to create algorithms that solve specified problems.

Structuring Teams for Project Execution

It takes large amounts of data, top-notch engineering, extensive business experience, and more to successfully create AI-enabled applications. It also takes a talented team. An AI team is usually put together at the beginning of a given project; in a federated system, this team draws talent from the CoE, IT, and the BUs.

There are a few things to remember when setting up a project team. It is important that AI scientists and data engineers work hand in hand with other engineers and business analysts for projects to run optimally. The best teams foster good relationships among these specialties. Without these relationships, AI initiatives can run into costly delays and produce degraded model performance. When data engineers assume the responsibility for such tasks as data preparation, AI scientists are freed to focus on what they should be doing – building and validating models.

Well-organized teams are also better able to deal with potential roadblocks. Many of these occur in the later stages of a project – problems such as tool incompatibilities or data governance, deployment, or security issues. Team members may also have difficulty getting end users to understand and utilize the predictions that their models produce. The best way to handle situations like these, of course, is to anticipate and deal with them before they arise.

A good team incorporates solutions to these issues at the beginning of a project, coordinating with stakeholders to make sure that project planning takes their needs into account; that expectations are realistic; and that appropriate resources are available. Doing this requires that a multiskilled, diverse team is in place from the beginning.

To set up a team for effective project execution, leverage the most reliable people in leadership roles. These are individuals who can bring together business acumen, AI and machine learning expertise, and IT capabilities, thereby enabling them to understand requirements and propose solutions. Leaders are usually either senior AI scientists who understand the needs of a business and have IT acumen, or experienced businesspeople with quantitative and data analytics experience. Some of the key roles required on AI project teams are described next.

- An *AI team manager* is responsible for utilizing her team's experience and expertise so that a project is as productive as possible. She should have the communication skills to connect with nontechnical personnel and enough technical expertise to communicate with her team members, to grasp what they are doing, and to support them in whatever ways are necessary.
- *AI scientists* are responsible for exploring the data and developing models and algorithms to build and deploy optimal intelligence solutions, assessing data for signal detection, and identifying approaches to AI that best fit the use case. They are the individuals who can understand business objectives for a use case and turn them into AI task objectives.
- *AI (or ML) engineers* specialize in building and deploying enterprise-scale platforms and intelligent products, productionizing proof of concept (PoC) models by combining software engineering and modeling skills. They deploy models on distributed machine learning and data frameworks (e.g. Tensorflow, Spark, etc.), and are responsible for the retraining, monitoring, and versioning of deployed models. Machine learning engineers are specialists in the infrastructure and operations of machine learning environments, and they work closely with data engineers in

supporting these processes. Some machine learning engineers are capable of prototyping end-user interfaces.

- *Data engineers* are responsible for dealing with the challenges of large-scale ingestion, storage, processing, and querying of data – enabling analytics by leveraging big-data technologies. Data engineers understand and design big-data architectures. They design and build data warehouses, data lakes, and data streams on distributed systems. They are responsible for maintaining data integrity, thereby enabling AI scientists to focus on the algorithms that will leverage this data.

- *Data analysts* measure, instrument, and iteratively explore data and develop data visualization and reporting tools to enable data-driven decision-making. They ensure that data is appropriately collected, and they are sometimes called upon to interpret analytic results to extract insights. They also help users understand how to employ existing models.

- The *AI DevOps* staff are usually part of a company's IT organization. Most IT organizations already have existing DevOps teams; they can be trained on ML or model ops, or other team members with appropriate MLOps experience can be hired. It is vital to ensure that they all get any training they need to support the AI engineers or data engineers. They are responsible for rapid development, testing, and release of the software, overseeing automated continuous integration and continuous deployment (CI/CD) pipelines, supporting bug fixes, and troubleshooting.

- *Technical project managers* bridge the gaps across the team. They need not be expert in IT, but they must have technical expertise as well as the political acumen to maintain good working relationships with each function at the table, as well as the management skills to ensure that projects adhere to the plan.

Although these are some of the more critical roles on AI projects, they are not the only functions necessary to make up a full team. Similar to other engineering projects, AI projects are a collaboration across the team along with business analysts, full-stack developers and engineers, product managers, SCRUM masters, solution architects, business intelligence analysts, data architects, and data-visualization engineers. Most organizations have roles similar to these in their day-to-day IT projects, so we do not cover them here.

Managing Talent and Hiring

When asked to define what an AI scientist is, many companies describe a person who has mastered machine learning skills, has enough business insight to understand exactly which use cases are the best ones to pursue, possesses the communication skills necessary to motivate the team, has the data engineering knowledge to be able to build robust data pipelines, possesses enough DevOps knowledge to set up and maintain infrastructure, has the talents of a developer who can prototype end-user applications, and more. But it is unlikely you will find all these skills in one person.

Companies should, instead, focus on building teams whose skill sets overlap and cover the breadth. Teams with blended skills bring a variety of useful perspectives to the table. Hiring an AI scientist who is part data engineer, a developer who is familiar with data engineering, or a data analyst with business experience is easier to do, and individuals who possess multiple skillsets often work together more effectively, since their expertise in more than one area makes them good collaborators. It can be useful, for example, if an AI scientist not only knows technical programming and mathematical modeling but also has some understanding of the business in question, so she can articulate the modeling decisions she makes as potential business trade-offs.

If the organization is unable to find such a well-rounded AI scientist, there are other solutions. A business might want to hire two people: one who can address algorithms and one who can address business strategy, since not many people have expertise in both. But each will need some experience in the other. Another alternative is to commit to training the new AI scientist in business strategy, a skillset the organization already has. Given the increasing demand for AI scientists, a business may have to hire one without much real-world experience and even some degree of discomfort about operating in a business environment. After all, they are experts in AI, not in business, IT, or deployment. Graduate schools teach AI theory and science, not how to put it into practice at businesses. Any assistance that the AI CoE or their counterparts in business units can give them should be appreciated. Companies might also want to instruct at least some of their existing data engineers in the art of AI. Many online and college courses lead to certification as an AI professional, including those from Microsoft, Coursera, and many universities.

Having said this, we need to acknowledge that, as noted in Chapter 7, sophisticated AI practitioners are very scarce. The number of open jobs is far larger than the available talent pool. This scarcity means not only that salaries are high but that AI scientists have many options from which to choose. The good news is that they are looking for the same things that most high-level, talented people are. They want meaningful work with a variety of use cases on which they can put their talents to bear and a variety of problems to solve. They want an opportunity to pursue their own research and the ability to suggest and implement new projects. They want a working environment that provides them with the things they need to succeed, including access to lots of data, and they do not want to have to spend the majority of their time on data cleansing before they can begin their work. They need access to a self-service AI platform so that they can operate independently and do not have to depend on IT departments to set up environments or procurement staff to source software.

To hire someone who meets the necessary criteria, organizations should let candidates know that they will be working in a creative and collaborative environment with a lot of other smart people, that the company is serious about its mission to create an AI program, and that there will be adequate budgets for projects as well as the full support of top management. Companies should also be clear about how their organizations differ from the competition. For example, if the opportunities within the company do not support what the AI person wants to do next – such as moving to applied research or working on a new product – the person will likely look elsewhere. A good AI scientist receives many offers; companies need to articulate what sets them apart in order to close the deal.

Unfortunately, the shortage of AI scientists is not the only problem facing organizations today. It has also become more challenging to hire data engineers – those people responsible for providing the AI scientists and others in an organization with the clean, reliable data they need. A search done on Glassdoor found that there were four times as many data engineering jobs open as there were job postings for AI scientists, and starting pay often began at $125,000 or more a year. Companies should be prepared for an active search process.

Sidebar: Careers in AI

More colleges and universities are offering majors in artificial intelligence. There are the usual suspects, such as Carnegie Mellon, Stanford, MIT, Columbia, and Harvard, but state schools such as Colorado State University and Eastern Michigan University also have well-regarded programs. AI-trained professionals can pursue careers as machine learning engineers, data scientists, research scientists, computer-vision engineers, and business intelligence developers. Many companies have already hired top academics to power their AI programs. In 2013, Google hired computer scientist Geoffrey Hinton to head its AI research effort. Facebook hired Yann LeCun, who works in the areas of machine learning and computer vision, to be its chief AI scientist. In 2015, Uber made off with nearly 50 researchers and scientists from Carnegie Mellon's National Robotics Engineering Center, including top robotic and automations specialist David Stager, now a lead-systems engineer[3] heading Uber's initiative to build self-driving cars. Yoshua Bengio is a professor at the Université de Montréal and an adviser to Microsoft; his Montréal Institute for Learning Algorithms (MILA) formed a partnership with IBM. There are many other examples. Although colleges expected to matriculate 400,000 computer science (not just AI) graduates by 2020, it is estimated that there will be 1.4 million jobs to fill by that time.[4]

Data Literacy, Experimentation, and Data-Driven Decisions

Because of the likelihood that AI will grow to be an even larger part of businesses, a healthy data culture is increasingly important. Having an AI or analytics capability within the organization without having the matching data culture to fully harness it diminishes its power. Executives and the AI CoE need to work together to plan for the transition to more data-driven decision-making within the organization. In today's transition period, there are three aspects to data culture that are worth noting.

Data literacy: The most comprehensive and up-to-date information in the world is not valuable to an employee who cannot read and utilize it. There needs to be a plan and tools in place to train employees to be able to utilize the data that will become a more ubiquitous part of their jobs. Increased data literacy also changes the cultural leanings of those who acquire it. In addition to training, a consistent data marketplace (see Chapter 9) needs to be enabled for business users. When business users, AI scientists, and IT engineers all use the same data marketplace and data catalog, they begin to speak the same data language. Another way to encourage this is simply to promote the utilization of more data in regular meetings. Illustrate for attendees the areas in which better decisions were made because they relied on data.

Experimentation: As mentioned before, in AI, iteration is an essential component of success. A test-and-learn approach to experiments helps to find the best use cases, datasets, and predictive models. Frequent experimentation also fosters a more data-centric corporate culture. The most successful data-driven companies are unafraid to make "mistakes"; they are constantly testing ideas. Managing an AI team is more like overseeing numerous experiments rather than merely providing straightforward assignments where expected results are known. But experiments need to be done with rigor. Experimenting does not just mean to try something to see if it works or not. It requires well-thought-out experiment designs, and control groups and test groups in place for learning from the hypotheses. The main driver for moving toward an experimentation culture is to create new knowledge. The more experiments a company can do, the more they are a learning culture. And by "learning," we do not mean getting trained on existing knowledge; we mean creating and utilizing new knowledge, from experiments. "Failure" is not only an option, but it is necessary for learning. A recent study by Robert Wilson, an assistant professor of psychology and cognitive science at the University of Arizona, and others, found that computers learned tasks the fastest when they were right 85% of the time and failed 15% of the time.[5]

Being successful at experimentation requires a mind shift in how experiments are viewed. It may not take 9,999 tries to find success, but it is doubtful that a first attempt will find the optimal solution; employees need to become accustomed to this. An experiment is a hypothesis that may be shown to be true or false. If there is no chance of it being false, it is not an experiment. Consider the example of a company in which the management team decided to change its culture to accept experimentation. It was challenging. Management itself had to learn how to talk and behave differently, as well as change how they held employees accountable. It took time. But it was ultimately successful. Executives stopped talking about failure and started talking about hypotheses being validated or invalidated, and invalidated hypotheses were looked on not as failures but as aids toward shaping future experiments. This company understood that not allowing people to fail meant not allowing people to experiment and succeed, and that without experimentation, nothing new would be developed – including successful new AI algorithms.

Data-driven decisions: A mature data-driven culture can be found in companies where data and algorithms support decisions. This does not mean that the company generates stacks of reports every day or just uses business intelligence tools. The key to creating a data-driven culture is to improve intuition with data and models. Scientists refer to making decisions in this way as evidence-based decisioning. However, they do not rely on data alone. The data needs to be combined with intuition, creativity, flexibility, personal experience, and a neutral perspective in order to make it fully valuable to companies.

Data literacy and experimentation are necessary to move toward more meaningful data-driven decisions across a company. In addition, users need to be trained on and comfortable with making decisions under uncertainty, because most data-driven models are probabilistic. To help with this, the AI CoE and management team can plan out a series of interventions to encourage more data-driven decisions. These can start with management asking business teams to explain the data showing up on reports presented to them (postdecision), then progress toward asking teams to explain why a decision should go a certain way based on multiple alternate data-driven models (predecision). Finally, the conversation can move to questions of optimization – questions such as, of all possible models, why is a certain option the best? We have seen these three stages be successful in moving data-literate companies toward data-driven decisions.

Conclusion

In this part of the book, we have covered how to develop an enterprise-wide AI strategy and how to bring it to life with the appropriate people, processes, and technology. In Part IV of the book, we will cover further details about the process and the business decisions involved in it with a hands-on example of a model and go into more details of the subcomponents for the architecture and how these work in different solutions patterns.

Notes

1. Battery (August 22, 2017). Getting Around "Moore's Wall": Databricks CEO Ali Ghodsi Strives to Make AI More Accessible to the Fortune 2000. https://www.battery.com/powered/databricks_getting_around_moores_wall/ (accessed September 30, 2019).

2. CNBC (July 11, 2019). Amazon Plans to Spend $700 Million to Retrain a Third of Its US Workforce in New Skills. https://www.cnbc.com/2019/07/11/amazon-plans-to-spend-700-million-to-retrain-a-third-of-its-workforce-in-new-skills-wsj.html (accessed September 30, 2019).

3. *New York Times* (September 11, 2015). Uber Would Like to Buy Your Robotics Department. https://www.nytimes.com/2015/09/13/magazine/uber-would-like-to-buy-your-robotics-department.html (accessed September 30, 2019).

4. Tech.Co (March 13, 2014). 1.4M Computing Jobs in America By 2020, But Fewer Computer Science Graduates. https://tech.co/news/computing-jobs-computer-science-grads-2014-03 (accessed September 30, 2019).

5. Nature Communications (November 5, 2019). The Eighty Five Percent Rule for Optimal Learning. https://www.nature.com/articles/s41467-019-12552-4 (accessed December 16, 2019).

Part IV

Delving Deeper into AI Architecture and Modeling

Chapter 12

Architecture and
Technical Patterns

The new spring in AI is the most significant development in comput-
ing in my lifetime. Every month, there are stunning new applica-
tions and transformative new techniques. But such powerful tools
also bring with them new questions and responsibilities.
Sergey Brin, co-founder of Alphabet

This chapter covers the technical architecture for the AI platform,
expanding on the high-level description from Chapter 9 and going
deeper into the subcomponents. To understand how the platform
works, we will look in more detail at the four core layers we reviewed
there as well as the elements within each layer. These layers are a data
minder for data management, a model maker for model experimentation and validation, an inference activator for deployment and
model serving, and a performance manager for ongoing production
monitoring. These components support the AI lifecycle discussed
in Chapter 8. We will also discuss design patterns for how to use
the platform in various solution scenarios, including for chatbots
and intelligent virtual assistants, personalization and recommendation engines, anomaly detection, physical IoT devices, and a digital
workforce.

AI Platform Architecture

Rather than assembling an AI platform from the ground up, it is common to use commercially available, cloud-based, machine learning base platforms developed by reputable software companies. Microsoft Azure, Amazon AWS, and Google Cloud provide these base platforms, among others. However, much of this commercially available software does not have all the components that AI scientists require during the AI lifecycle. If these components are lacking, companies can fill in the gaps by either building the missing pieces in-house or acquiring them as open-source tools or commercially available applications that can be integrated into the AI base platform.

Data Minder

The data minder is used to manage the integration and data of a platform. This component is where data is gathered and cleaned and where appropriate access to that data is controlled. This layer includes data lakes, databases, data files, and data warehouses, with inputs both from information sources across the company and externally.

Data sources are the primary locations from which data is collected first hand. These could include the organization's customer relationship management (CRM) system, the enterprise resource planning (ERP) system, the order management system (OMS), logs from the customer website or mobile app, Internet of Things (IoT) devices or sensors, still or video cameras, wearables, or any other sources. Some of the data and integration layer may even be externally focused: that is, it ingests data from sources that are outside the enterprise. These can come in through a batch or streaming pipeline.

The raw *data store* is where all incoming data is stored without modification, whether the data is structured, which might mean it comes from a database table, including time series data, or unstructured, such as an image file from a drone or satellite, a video, or a

Word document. The raw data store is meant to house data from the source system "as is," but validations are required to ensure it is consistent with data from the source system. These validations can include structure and format validation and validation of source and destination record counts or data distribution and profiling. Invalid data is flagged and stored for further analysis and correction.

Data pipelines are created to flow data from different origins to various destinations. The data is then further processed – undergoing sanitization and filtering, standardization, normalization, referential integrity checks, and other types of curation – to make the data usable for downstream analytics and AI. This data is then stored the *curated data* store. Some data wrangling or business logic–based transformation is performed at this stage, as well making downstream processes more efficient.

Data labs are where AI scientists pull in their data to use throughout the experimentation and modeling process. They then make any necessary changes, including transformations, bias-reduction, scaling, or other data preparation steps needed before or during modeling. Data labs can be thought of as the development environments for the modelers, and there is at least one person, or a small team of people, working in each lab.

The *data governance* component manages the data governance and stewardship workflow and leverages the *data quality* component to understand the quality of the data and determine what actions are required to ensure that the uses of the data are consistent with corporate policies and applicable laws and regulations.

The *data synthesizer* component is used to do two things. First, it synthesizes new data for model experimentation when actual information is not yet available. Synthetic data is data that is generated programmatically, as opposed to real-world data, which is collected. Second, it synthesizes data where there are privacy concerns about existing data. Synthetic data is generated with precisely controlled statistical distributions that reflect real-world scenarios. Often, the artificial data model is set up to mimic a real-data model. AI models trained on the synthetic data can eventually be retrained on the real-world data after they are collected.

The *data labeler* component is also used to label existing data-sets to prepare them for supervised learning. The label is the target attribute that the machine learning algorithm tries to predict. This component manages a workflow for human labeling, but it may also have an automated labeling component. It serves the features to a user who then enters a label (free form or from a predetermined set). This labeling can be done with structured data, audio, images, and other data types. For example, a human labeler might be asked to draw boundaries around various objects in an image.

There are two data lakes for storing data from processing. The *features data lake* stores any computed features that were developed from the curated data, and that can be reused. These include embeddings created for categorical data such as customers or products. Embeddings are mathematical representations of categorical items, such as retail products, that convert each item and its properties into a numerical vector. The *insights data lake* stores any computed model results from running the models in batch or real time. Users usually see the batch insights through reports or visualizations. Results from real-time model outputs are returned through the API layer when the model is called through an API, but these results are also stored in the insights data lake for model-performance monitoring.

The *knowledge graph* is a graph of entities, relationships, and other information gathered from unstructured text data such as Word documents and PDF files. It is used to more efficiently and precisely locate information that was in the documents. Often, there is more than one knowledge graph, depending on the purpose for which it was built – for example, to find answers asked of a virtual assistant about the expertise of people in a company or about company policies.

The *data marketplace* is a comprehensive view of all the data and features available to AI scientists (and other users). This information is easily browsable to enable users to understand what data is available and includes descriptions of the data, associated metadata, information on the relationships among the data elements, data lineage, and so on. In many instances, available information also includes data profiles such as counts and distribution of the rows of data. AI scientists as well as business analysts, business users, and business intelligence (BI) developers use the data marketplace to understand all the data that is available in the company. This data marketplace is key to enabling data literacy and a culture of data-driven decisions within a company.

Model Maker

The model maker, also known as the experimentation layer, is where AI scientists develop, validate, and iterate their hypotheses. This layer needs to support them through training potentially hundreds of models before they arrive at the ideal one for a specific use case. It should also help them through the model risk assessment for each model and in developing any necessary corrections. The model maker has three groups of components for the AI scientists to use (shown as curly brackets in the model-maker component in Figure 12.1) – for working with the data, for making the model, and for model assurance.

The data components in this layer are for AI scientists to use so they can understand their datasets before they can move on to the modeling steps. The *data-visualization* component is a set of libraries that allow the user to easily create various visualizations. The *data preparation* component is a set of libraries that help with data imputation for missing data and convert data to the right format for modeling. The *exploratory data analyzer* helps discover patterns, find anomalies, look for bias in the data, and understand various statistical summaries of the data. The *signal explorer* finds correlations in a dataset when it is given a data lake and a target data column to predict. It looks for which other features are correlated and are likely predictors for this target column. Other than time, the advantage of an automatic approach over a manual one is the ability to find unexpected patterns more easily.

AI scientists use modeling components to develop their models. The *algorithm frameworks* include various types of algorithms that can be trained on the data to create models. Frameworks such as

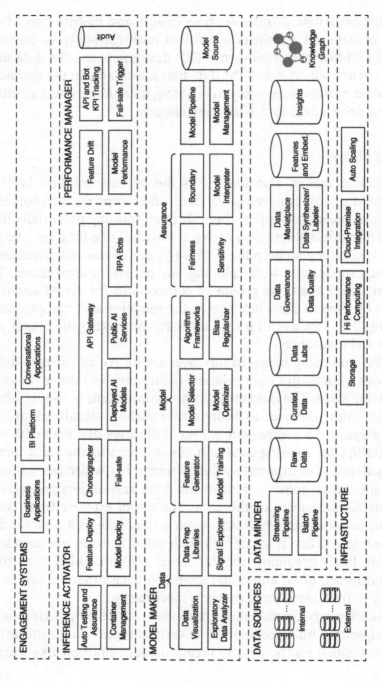

Figure 12.1 Architecture components for an AI platform.

scikit-learn, TensorFlow, PyTorch, and others are set up and managed here. These frameworks include algorithms that cover simulation and optimization, anomaly detection, pattern recognition, prediction, knowledge mining, and potentially others.

The *feature generator* creates and evaluates derived features generated by combining existing data features. For example, if there are two base features called start-date and end-date, the feature generator creates a feature (duration) that is the difference between the two. It then determines whether this derived feature is predictive of the target or not. The feature generator is similar to the signal explorer, but it generates and evaluates derived features rather than just working on the base features.

Model training includes different ways to configure the models and the different types of learning that can be leveraged. Modeling includes rules configuration for RPA and inference engines, batch training for machine learning and deep learning models, online training using reinforcement learning, distributed (or federated) learning for edge devices, or combinations of these. The *bias regularizer* is used in the modeling process during training to penalize models for high bias and low accuracy. This process ensures that bias is managed and that the model does not simply optimize for predictive power irrespective of bias. It is part of the model risk-governance steps discussed in Chapter 10. The *model selector* helps compare the performance of the different experiments to select the optimal one. The *model optimizer* helps tune the AI models' hyperparameters for the best performance. It runs the model training and evaluation multiple times. It also tracks the metadata and uses optimization methods to find the optimal hyperparameter settings for model performance. Without this component, a great deal of time may be spent tuning hyperparameters manually, and this can make it challenging to find the best performing parameters.

The *model management* component helps track models and changes to them to document which model experiments have previously been conducted and how to reproduce the results of an experiment accurately. This component uses configuration management tools to track such things as what features were tested and discarded, what modifications were made to data pipelines, and what compute resources were made available to support sufficient training. Together with configuration information, model management accelerates the consistent deployment of AI services while helping to reduce redundant work. The *model source* is the repository for all model code. The *model pipeline* is the component where the configurations for the model pipeline, from data to modeling to deployment, are managed.

Inference Activator

The inference activator deploys models into production and powers AI during run-time inferencing. The models are usually invoked through an API call from business applications using data that was not part of the model training and testing. Then the models are executed to make predictions. The continuous integration and continuous deployment (CI/CD) DevOps pipeline houses the tools that manage model deployment, including automated testing and model assurance. The *testing and assurance* component executes automated testing. This testing includes model validation to ensure model performance is as expected in each new environment; end-to-end testing to validate that the data pipeline is accessible and APIs for the model are working; and model risk tests, such as fairness, boundary condition, and sensitivity testing, that were discussed in Chapter 10.

The *model deployment* component packages and deploys the AI models, usually in containers to be used through APIs for real-time use, or within a data pipeline for use in batch mode. Containers are an open standard for packaging and distributing applications, including all software dependencies, so that the model can run quickly and reliably in any environment. They are in everyday use today for both model deployment and in deploying and scaling other kinds of software. The *feature deployment* component deploys code for any feature transformations required on the input data. Sometimes this is integrated into the model API, but sometimes it is separate from it.

For AI in run time, there is an *API gateway* that manages the set of APIs through which applications can invoke or trigger pre-trained AI models. These include *deployed AI models* that were built on the platform and *public AI services* that are available from the large cloud providers, such as Microsoft Cognitive Services, a variety of startups, and many smaller companies. Lastly, *RPA bots* can also be accessed through the APIs as needed. The APIs are called by connected business applications that consume the model or from business intelligence (BI) platforms to gather and present information in reports and dashboards. These applications usually have a human user. The APIs can also be enabled for different types of machines, consumption, such as autonomous systems that are synchronous or asynchronous but that have no human consumer. Sometimes the user interface of the application itself requires AI APIs, such as requests from conversational user interfaces in virtual assistants or for gesture control, eye-tracking, biometric recognition, and ethomic (movement-based) interfaces.

The *choreographer* is configured when multiple models need to work in concert with each other. Sometimes this is handled inside business applications in the top layer; at other times, it is better to allow the model interactions to remain in the inference layer, depending on the specific use case. Some "intelligent products" are composite choreographs of multiple lower-level APIs – for example, sentiment analysis from a voice file may include choreography among a speech-to-text API and various natural language processing (NLP) APIs.

The *failsafe* component allows administrators to configure the failsafe mechanism for models for which such mechanisms have been deemed necessary. The *container management* component helps manage the deployed containers behind the model APIs and orchestrates the storage, networking, and computing infrastructure to ensure the application is available. This is where configurations are managed to scale the number of containers for a model based on CPU or GPU usage, accomplishing such tasks as restarting or replacing containers that fail.

Performance Manager

This performance manager monitors run-time AI models – models that are in active use. The *feature drift* component follows the data distribution of all the input data for each feature and compares it to the distribution known at the time of model training. If these distributions are sufficiently different (that is, they exceed some differential threshold), then the model may have to be retrained. Drift is a leading indicator of any potential issues in model accuracy. The *model performance* component monitors whether a model's output and predictions are still meeting a defined threshold. For example, is the model accuracy still at or above 90%? Sometimes, there is a time lag between when the model predicts something and when the predicted event occurs, so model performance is often a lagging indicator of a model's fitness for use.

What we have covered so far are all the components and subcomponents of a robust AI platform. You can think of this as the static view of the architecture. Next we will cover the dynamic view.

Technical Patterns

This section will delve more deeply into a variety of technical patterns for some common applications of AI and machine learning, as well as how these applications may be implemented in a production environment. These technical patterns describing how the platform is utilized often come up in many use cases. You can think of them as the dynamic view of the architecture.

Intelligent Virtual Assistant

An intelligent virtual assistant is software that has a conversational or message-based user interface that supports customers and employees in doing various tasks. Common examples include Apple's Siri and Amazon's Alexa. More are being developed by businesses to help employees to do such things as getting immediate answers to investment questions, and to support customers to do things like getting conversational access to bank accounts and provide advice.

To work well, virtual assistants require a few core components. The first is the ability to understand a user's command or question. If the input comes from voice, then the voice command needs to be translated to text through a speech-to-text model. Then natural language processing needs to be applied to the text to understand the speaker's intent. For example, if a speaker asks "What is the stock price of Microsoft?," the virtual assistant must be able to interpret the user's intent – stock price – and the entity or entities associated with the intention – Microsoft.

The complexity of understanding a question increases based on the user's request. The question can have multiple entities, such as "What are the stock prices of Microsoft and Google?" or it can have a compound entity such as "the technology sector." The request can also have multiple intents, such as "When is the next available flight to Chicago, and what is the weather there?"

Once the system understands the request, the next step is to determine a response or provide an answer. If the request is a command such as "Turn the volume down," then an API can be called to act on it. If it is a question that requires an answer, it can be handled in multiple ways. For structured data lookup, such as the stock price of Google, an API call or a database search might provide an adequate solution. If the answer lies in unstructured data, then it needs to come from a knowledge modeling solution. An example of a knowledge model is a knowledge graph that structures the text from documents. You can see the results from a knowledge graph if you search for something like "What is Barack Obama's height?" Although the Google search engine provides links to documents containing this question or keywords from it, at the top of the page, it also gives the answer from its knowledge graph.

Knowledge modeling goes through multiple steps (see Figure 12.2) to develop a machine-readable knowledge repository. This process can include breaking up the original document into smaller

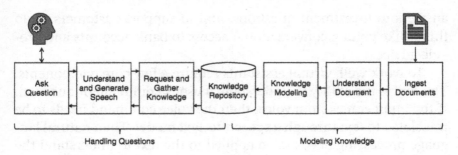

Figure 12.2 Question-and-answer systems built on knowledge modeling.

segments, sometimes called "answering units"; extracting topics of discussion and tagging the section with those topics; extracting entities and concepts from each sentence in the document; and building a knowledge graph of these entities and concepts and their relationships. Using this approach, you can find more precise answers in the knowledge graph or the answering units.

Lastly, to make the virtual assistant work, the users' interactions need to be managed. This function includes asking clarifying questions, confirming what the virtual assistant is going to do to handle a request, and deciding when to speak or respond or how to present information back. In some cases, an assistant needs to help users through a guided conversation – for example, if the customer is applying for a mortgage loan. In these cases, the virtual assistant also needs to know what steps have already been taken and what needs to happen next.

The goal of the intelligent virtual assistant pattern is to interact with the user naturally, using natural language through voice or text and touch. In more recent applications, companies are also incorporating gesture, augmented reality, and eye tracking as modes of input. Given that users might ask any type of question in a wide variety of ways, companies need to build solutions in which the NLP training can happen quickly on an ongoing basis without having to manually craft the knowledge modeling. Active learning by the system should be incorporated into the solution so that it can be quickly trained on any questions it does not yet understand.

Personalization and Recommendation Engines

Personalization is about creating an individualized experience for every user with whom a business is interacting, usually digitally. In the past, this was just about delivering the right item to the right person, but today, much more is possible. Now it is commonly described as providing the right item (message, content, or product) to the right person at the right time using the right channel.

To make this kind of hyperpersonalization work, companies need to take a few steps. First, various types of information are aggregated for each customer so that insights can be extracted to understand that customer in more fruitful ways. This data may include profile and demographics, historical transaction data, data from the customer's use of the company's mobile apps or websites, impressions data from ads, and potentially other kinds of third-party data. All this data needs to be stitched together using the customer IDs. This type of data store is called a *customer data platform* (CDP).

Next, multiple AI models are developed and combined to improve the business's understanding of each customer. These models might include predictions about which products consumers would likely purchase. What is their sensitivity to the product price? Is there an alternative product they would prefer? Is there an additional product they would like, given what they are purchasing? What channel do they respond to best – email, ads, messages? When is the best time to reach them, and when are they most likely to respond positively? What language in the ad or email leads to the best response rates? These models are used in concert; for example, depending on the product to recommend (from model 1), the optimal channel (model 2), the time of day (model 3), and the language to use in the message (model 4), the appropriate message is shown to the user. Figure 12.3 shows a sample set of models (black boxes) in a retail business being used to create the right interaction with the customer (gray boxes at the top).

Finally, interactions with the customers are orchestrated through the most relevant touchpoints through customer journey orchestration applications. The AI models generate insights

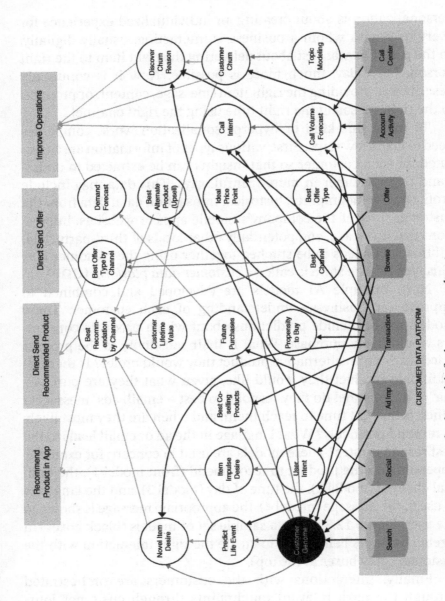

Figure 12.3 Leveraging multiple models for hyperpersonalization.

and make recommendations for the most appropriate next steps. The orchestrator takes action on these by receiving the insight or recommendation and triggering the proper response in another system (see Figure 12.4). These can include such things as sending an email to the customer with an offer or showing product recommendations when the customer is browsing the company's website or app. Because there are many models being exposed through APIs and many customer journey touchpoints to orchestrate, not using a journey orchestrator creates hundreds of point-to-point connections that become impossible to maintain in the long term. For this reason, a journey orchestration solution is ideal.

The most common way to build recommendations is with batch processing. However, this has its disadvantages. One is that data can become stale between sessions. Another is that the recommendations may be based on the customer's entire history rather than just their current session, which can indicate what they are in the mood for at the moment. For these reasons, more companies are building real-time recommendations engines. The batch versus real-time question is covered in Chapter 9.

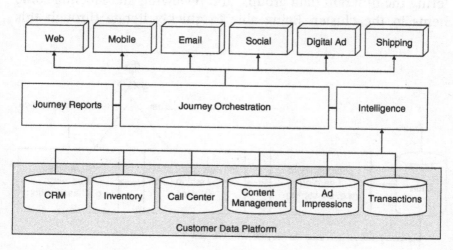

Figure 12.4   Orchestrating personalization interactions.

Anomaly Detection

Anomaly detection is regularly used to identify particular data that differs significantly from the rest of the data in a dataset. Detecting fraud, diagnosing cancerous tumors, performing risk analysis, and identifying patterns in the data to facilitate insights are only a few of the applications in which anomaly detection is used. To be able to find anomalies, it is necessary to have a specific and detailed description of what "normal" data looks like. Then, anomaly scores are computed for each new data sample, by comparing this sample with the model norm. If the deviation exceeds a predefined threshold, the data sample is deemed to be an outlier or anomaly.

Detecting anomalies starts as an unsupervised task, because the anomalies arise from unknown data patterns, so there is no labeled data to learn from. Unsupervised anomaly detection, however, often fails to meet required detection *rates* in many business use cases. In those cases, labeled data is required to refine the model. This can be critical; in some instances, such as cancer detection, highly accurate models can mean the difference between life and death.

The crux of successful anomaly detection lies in using a workflow (see Figure 12.5) that starts with unsupervised learning, clustering the different data groups, and reviewing and labeling some items in the cluster. Being able to quickly iterate through this

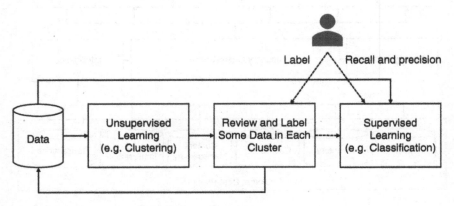

Figure 12.5 Activities for anomaly detection.

process provides enough labeled data to enable supervised learning for previously identified and labeled anomaly patterns, while the unsupervised process continues to find new anomaly patterns. Platforms that allow for rapid data labeling, experimentation, and model deployment – ideally autonomously – are critical to scaling anomaly detection successfully.

One of the goals in anomaly detection is to avoid false alarms while identifying as many legitimate anomalies as possible. To determine detection accuracy, we utilize two measures: *recall* (the number of true anomalous points predicted divided by the total number of true anomalous points – both true positive and false negative) and *precision* (the number of true anomalous points predicted divided by the number of anomalous points predicted – both true positive and false positive). Both recall and precision measurements need to be calibrated accurately over time to detect anomalies effectively. To extend this to work for real-time anomaly detection, we create an "anomaly score" that shows the "distance" (in data space) of incoming data from standard as well as anomalous data. Using this distance, incoming data can be prelabeled anomalous or normal – for example, in detecting credit card fraud at the time of transaction.

Ambient Sensing and Physical Control

There are many current examples of ambient sensing, an increasing number of which control physical systems. In manufacturing, we can use this type of capability for health and safety scenarios that can include keeping a person safe by flagging when she is getting too close to an unsafe condition, or making the problem of losing things less burdensome by flagging where a lost item might be with instructions on how to find it. But many applications go beyond industrial settings. This capability makes it possible to create several exciting retail experiences, including grab-and-go shopping and immediate and on-the-spot assistance when needed. There are many compelling scenarios enabled by this pattern that are just emerging.

The steps businesses need to take when implementing AI for ambient sensing and physical control are more involved than those required to set up AI systems for other companies. That is because a much tighter integration is needed between the digital and physical aspects of the operations. This pattern covers the use of machine learning and other AI approaches in which the system interacts with the physical world in some way, utilizing the Internet of Things (IOT). The IoT is the collection of devices containing electronics, sensors, actuators, and software that connect, collect, and exchange data. Home automation devices, including accessories such as the Nest thermostat, are an excellent example of IoT.

IoT systems consist of *edge devices* – devices that are sitting closer to the machines (e.g. a robot in a car manufacturing plant) rather than close to the servers in the core network of the company. These devices have sensors that detect temperature, humidity, pressure, gas, light, sound, radio frequency identification (RFID), which enables information stored on a tag attached to an object to be read, and near-field communication (NFC) that allows smart devices to communicate with other electronic devices. Other devices such as ultrasonic sensors, flow meters, and cameras may also be used.

The information flow may be one-way, such as an intelligent camera that does facial recognition on-site (at the edge of the network), or it may be two-way, such as an IoT thermostat device that both measures the ambient temperature and adjusts the thermostat accordingly. Edge devices can also include actuators to control machinery. These actuators work in conjunction with switches, relays, industrial computers adapted for use in the control of manufacturing processes known as programmable logic controllers (PLCs), and motors. Sometimes, edge devices act as both sensors *and* actuators.

The most common use of this pattern is to enable a system to accomplish a task, achieve a goal, or interact with its surroundings with minimal to no human involvement. This pattern may be used to autonomously control a system's hardware, thereby minimizing human labor, or to quickly make decisions or adjustments in systems that need intelligence where human intervention would be too slow. Smart thermostats, autonomous vehicles, automated machinery within manufacturing plants, and other machines and robots are examples of the use of this pattern, and a host of additional so-called smart products are already being manufactured, from the mundane (e.g. pet feeders) to the exotic (e.g. human exoskeletons). More devices, large and small, are joining the IoT world every year.

A critical part of this pattern is to determine the nature of an object when supplied with some form of structured or unstructured data. This data could include images, video, audio, or IoT sensor data; the goal is to have some aspect of the data labeled and tagged by identifying, recognizing, or classifying it. Use cases include object and image recognition, facial recognition, sound and audio recognition, handwriting and text recognition, gesture detection, and any classifying of behavior that is motion based.

The overall IoT architecture consists of three layers (see Figure 12.6): the edge layer, the IoT services layer, and the enterprise layer. The *edge layer* consists of edge devices, such as sensors, and the *IoT gateway*: the device or software program that connects the cloud to controllers, sensors, and intelligent machines. Devices may be distributed in a wide variety of locations depending on the use case, which limits their communication to a relatively short-range network because of connectivity, processing, and power limitations.

The IoT gateway comes into play when devices need to communicate with the rest of the world. The IoT gateway usually contains a data store for IoT device data, one or more services to analyze data coming either directly from the devices or from the data store, and control actions based on the incoming data. The IoT layer also might have edge intelligence with different levels of processing capabilities, allowing machine learning models to draw inferences if there is enough computing power available.

Data from the edge layer is consolidated, processed, and analyzed by the IoT *services layer*. This layer also manages devices, physical assets, and processes and forwards control commands from the *enterprise layer* to the edge layer. Data from the edge and IoT services layers is received by the enterprise layer, which is the

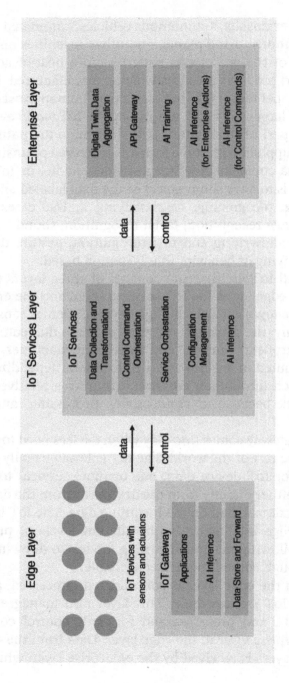

Figure 12.6 Interaction pattern for IoT and edge devices.

layer we have focused on in previous chapters. Trained models in the enterprise layer then trigger reactions to new sensor data coming in and send actions back to the actuators, transmitting simple triggers such as alerts. More complex triggers include issuing control commands to the other two layers modifying main or subsystem instructions, tasks, or processes.

Machine learning models are most often trained at the enterprise layer, once the data has been aggregated there. Model inferencing can also happen in this layer, as described earlier, or the model can be installed on the edge device to respond in the local environment. A new approach, called *federated learning*, is being explored currently for training ML models across edge devices without bringing all the data back to a central location. In the enterprise layer, it is critical to ensure the metadata remains accurate. Knowing, for example, which sensors are connected to what component, which components are part of the same machine, and which machines are part of the same plant or factory is necessary to develop a digital twin.

Digital Workforce

This pattern uses a combination of robotic process automation (RPA) and machine learning to create a virtual workforce. These virtual employees follow predefined and documented processes to the letter, without error, omission, or deviation. Being able to leverage this kind of digital workforce reduces operational costs and rework and enables the creation of workers on demand as workloads change due to demand or supply changes caused by seasonality, weather events, or other similar effects.

The digital workforce is made up of bots that perform certain business activities to automate a business process. After logging into enterprise IT systems much as a user would, using virtual desktops, these bots can execute multiple parallel business processes simultaneously. There are generally three types of work that digital workforce bots do: action automation, analysis automation, and decision automation.

Action automation is sometimes referred to as "fingers on keys." Bots performing action automation can connect to other software systems such as an enterprise resource planning (ERP) system, usually through the front-end screen. These bots focus on high volume, repetitive, and rules-based activities that are digitally triggered, such as taking actions when an invoice file is received. This might occur during the close of business-day activities for a commodity trading organization, for example, when bots can be utilized to execute and monitor a series of processes across multiple systems like valuations, simulations, report generation, and data reconciliation.

Analysis automation is the gathering and interpretation of information, such as natural language free-form comments in a transaction system. Bots can ingest market data from different nonstandardized sources in a structured format (e.g. data from the web or feeds) or unstructured data (e.g. brokers' emails, Word documents, PDF attachments). The bots then extract and transform data into standard information templates and key them into downstream systems. In some cases, a bot may call an AI API for image recognition, optical character recognition (OCR), or natural language processing as needed.

Decision automation is the automation of a decision on behalf of a user based on some defined policies, or the ability of a bot to make a recommendation to a user who can then make the final decision. An example is when an invoice is received from a commodity trading counterparty as an email. The bot compares the invoiced amount to the amount in its trading system. If the difference in value is less than a specific dollar amount (set by policy), it creates a cash adjustment and marks the invoice for payment. Otherwise, it leaves the invoice for a human analyst to review. This is a trivial example of a "decision," but in more complex situations, the bot may also call a machine learning API to help make a decision, such as in a small loan application approval process.

Figure 12.7 shows a typical digital workforce architecture. A *bot server* stores the RPA bot configurations and the specific tasks the bot should execute in the systems. A *bot control room* manages and controls multiple bots to complete an automated business process that is associated with that group of bots. The control room assigns

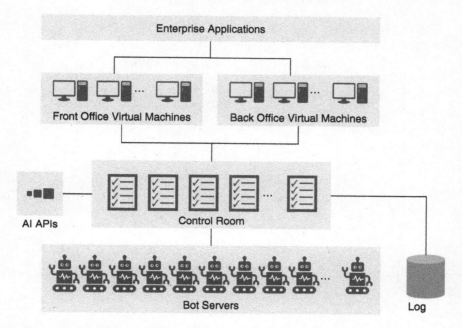

Figure 12.7   RPA-based digital workforce architecture.

the business process to be executed to a bot in a group based on bot availability, as well as on process schedules defined and configured in the control room. There are two types of bot triggers – front office and back office. Users trigger front office bots when they need them. The resulting information is then sent back to the control room. The control room itself triggers back-office bots, which also report status back to the control room.

Conclusion

Now that we have reviewed the structure of AI platform architecture, the next chapter will be devoted to illustrating how the model-building process works. We will do so by examining a specific machine learning application designed to understand customer churn in the telecom industry, but the same methods apply to other use cases.

Chapter 13
The AI Modeling Process

What is vital is to make anything about AI explainable, fair, secure and with lineage, meaning that anyone could very simply see how any application of AI developed and why.
          Ginni Rometty, chairman, president, and CEO, IBM

In this chapter, we will work through a simple example of a machine learning application to give readers a better understanding of how the model-building process works, expanding on the AI lifecycle discussed in Chapter 8. For our example use case, we will use a hypothetical telecom company that is setting up an AI model to solve a specific business problem of customer churn. We will go through a step-by-step explanation of how the modeling process works, why specific steps are essential, and how to avoid potential pitfalls. This chapter is not meant to show the reader how to become an AI modeling expert but, rather, to help managers and executives who will be supervising or interacting with their AI teams to better understand what these teams are doing and why, and what business decisions are involved in the modeling process.

Defining the Use Case and the AI Task

There are a few key questions to answer to define an AI use case well. The first is, what needs to be the output of an AI model or sequence of models – that is, what prediction must be made or what task must the model accomplish? The second is, what decision or activation within a business flow will the AI task enable so that it can be of value to the business? Too often, teams working on AI projects will only answer the first question and not the second, but it is activation that ensures AI projects will not be abandoned. The proliferation of discarded proofs of concept (PoCs) across hundreds of companies is evidence that the question of activation is not being asked early and often enough.

Businesses spend a lot of effort and money to acquire new customers by building and maintaining sales and marketing teams, paying for advertising, and giving individuals discounts or other incentives to become customers. Businesses within most industries therefore generally find it less expensive to retain customers than to acquire new ones. However, with a proliferation of brands offering excellent services and products, most companies are facing the challenge of customer churn.

Customer churn refers to customers who stop using a company's products or services. It is usually measured as a percentage in a specific time window, such as one year. Predicting how to retain customers turned out to be an essential early success of modern machine learning. Churn is an issue within our imaginary telecom company; executives want to figure out which groups of customers are likely to switch away to a competitor and then work out a way to retain these customers at a lower cost than it would take to acquire new customers. The telecom's AI task for this use case will, therefore, be to look at customer data and predict, with sufficiently high accuracy, the threshold at which their customers are likely to churn – say, in the next three months.

This problem can be treated as a classification problem (customers who will churn versus customers who will not), or as a regression problem (probability of churn for each customer). Keep in mind that this AI task (the model) will predict if the customer will churn or not. It will not recommend the best course of action for how to treat the customers who may potentially leave. The activation in our case will be to take the list of customers who are highly likely to churn and create marketing campaigns targeted to these customers, with either messaging or offers to make a case for them to stay. One example of this might be to offer these customers discounts on their next bill to incent them to remain with our telecom company, as an offer through direct email, or as an outbound call through the call center, or as a message on the billing and payment part of the telecom's app, where they can accept it with one click.

While implementing any use case for AI, a team often comes up with additional ideas for use cases – in our experience, an average of four new ones per use case. In our telecom scenario, additional use cases might include predicting what type of incentive each customer would be most responsive to (e.g. future discounted rate, free weekends, or messaging about being an environmentally friendly company). Another is to predict what channel of communication each customer would prefer, that is, which channel would increase the probability of positive responses (e.g. email, agent call, or mail). A third is to predict the customer lifetime value so that incentives are given only to those customers who will be valuable in the long run. On any project, new use cases like these should be added to the backlog of use cases as they are identified.

The next question that needs to be answered for a given use case is about value. What business value are we creating, or what business objective are we serving? It is crucial to make a high-level value identification during the use case definition, both for relative prioritization (see Chapter 8) and to ensure that there is actual value if the use case is developed. This helps to drive adoption of the activation. In our example, if it takes $200 of advertising and marketing effort to acquire a new telecom customer and the company has 10,000,000 customers with a 15% annual churn, reducing churn by two percentage points will retain 200,000 customers each year, leading to an annual savings of $40 million. Retaining customers may require an incentive of $50 per customer on average, which would make the cost of retention $10 million, resulting in a net savings of $30 million per year. This example makes many simplifying assumptions and is not meant to describe a business case in detail; for an actual business case determination, enough details should be provided to build a more robust case before substantial investments are made.

Selecting the Data Needed

It is not always clear at the outset of the AI modeling process what data should be collected to develop the model. It is a good idea to list potential reasons for churn and use these as indicators of data that should be obtained. For example, churn could be due to a customer's bills being too high, driving her to look for a better deal (i.e. get billing data) or because she was having problems with her service (i.e. get service outage data or data about the frequency of calls to the call center). It is also a best practice to not assume these are real reasons – they simply provide a way to brainstorm what data to collect. These can be tested as hypotheses using the data. Once you have the data, let it speak for itself.

In our example, we will use a publicly available sample dataset about customers at an anonymous telecom company.[1] It is likely that this data has come from multiple systems and was connected by customer identification. A good source for public datasets such as this is Kaggle. Our dataset has approximately 3,500 rows of data (observations), with each row representing one customer. The columns are as follows:

1. CUSTOMER_ID: an identification number for the customer
2. STATE: the US state the customer is in
3. AREA_CODE: customer's phone area code
4. PHONE_NUMBER: customer's phone number
5. ACCOUNT_LENGTH: duration of months the customer has had an account

6. INTL_PLAN: whether the customer has an international plan
7. VMAIL_PLAN: whether the customer has a voicemail plan
8. VMAIL_MSG: number of voicemail messages the customer has received in a given month
9. DAY_MINS: number of daytime minutes the customer has used in a given month
10. DAY_CALLS: number of calls the customer has made during the daytime in a given month
11. DAY_CHARGE: total dollar charge for daytime calls in a given month
12. EVE_MINS: number of evening minutes the customer has used in a given month
13. EVE_CALLS: number of calls the customer has made during evenings in a given month
14. EVE_CHARGE: total dollar charge for evening calls in a given month
15. NIGHT_MINS: number of nighttime minutes the customer has used in a given month
16. NIGHT_CALLS: number of calls the customer has made during the nighttime in a given month
17. NIGHT_CHARGE: total dollar charge for nighttime calls in a given month
18. INTL_MINS: number of international minutes the customer has used in a given month
19. INTL _CALLS: number of calls the customer has made internationally in a given month
20. INTL _CHARGE: total dollar charge for international calls in a given month
21. CUST_SERV_CALLS: number of customer service calls in a given month
22. CHURN: a yes or no column indicating whether the customer unsubscribed or not within three months of the period covered by the data in the other columns

Setting Up the Notebook Environment and Importing Data

We will develop our model using an open-source Jupyter notebook environment and the Python programming language, which is currently the most popular programming language for AI and data science. Many existing libraries can be used for this project – libraries for data manipulation, for matrix algebra, for graphing, and so on. We will use the scikit-learn (sklearn) open-source machine learning library. Figure 13.1 shows importing of various libraries from sklearn and also importing of other open source libraries such as SMOTE.

Utilizing code in Figure 13.2, we import data that was in a comma-separated values (CSV) file in a folder on the C drive. We do this by utilizing a Python DataFrame data structure, and we call

```
1  # import math and data libraries
2  import pandas as pd
3  import numpy as np
4  from scipy.stats import uniform, randint
5
6  # import visualization libraries
7  import matplotlib.pyplot as plt
8  import seaborn as sns
9  from mpl_toolkits.mplot3d import Axes3D
10
11 import missingno as msno
12
13 from imblearn.over_sampling import SMOTE
14
15 # import sklearn machine learning libraries
16 from sklearn.preprocessing import LabelBinarizer, label_binarize, Imputer, \
17          LabelEncoder, OneHotEncoder, StandardScaler
18
19 from sklearn.compose import ColumnTransformer
20 from sklearn.pipeline import Pipeline
21 from sklearn.impute import SimpleImputer
22
23 from sklearn.model_selection import train_test_split, cross_val_score, \
24          GridSearchCV, KFold, StratifiedKFold, RandomizedSearchCV
25
26 # import the necessary model types
27 from sklearn.linear_model import LogisticRegression, LinearRegression
28 from sklearn.naive_bayes import GaussianNB
29 from sklearn.svm import LinearSVC
30 from sklearn.ensemble import RandomForestClassifier
31 import xgboost as xgb
32
33 # import model performance tools
34 from sklearn import metrics
35 from sklearn.metrics import precision_recall_curve, roc_curve, auc, \
36          accuracy_score, make_scorer, recall_score, \
37          precision_score, confusion_matrix
```

Figure 13.1 Importing relevant libraries that will be used.

```
1 # set the folder and file names from where you want to get data
2 folderName = 'gdrive/My Drive/Colab Notebooks/Data/'
3 fileName = 'customer_churn.csv'
4
5 # create dataframe and read file into dataframe
6 imp_data = pd.read_csv(folderName + fileName)
7 imp_data.shape
```

(3333, 22)

Figure 13.2 Importing the data for customer churn.

```
1 imp_data.head()
```

	CUSTOMER_ID	STATE	AREA_CODE	PHONE_NUMBER	ACCOUNT_LENGTH	INTL_PLAN	VMAIL
0	10001	KS	415	382-4657	128	no	
1	10002	OH	415	371-7191	107	no	
2	10003	NJ	415	358-1921	137	no	
3	10004	OH	408	375-9999	84	yes	
4	10005	OK	415	330-6626	75	yes	

5 rows × 22 columns

Figure 13.3 Looking at the top few rows of the data.

it imp_data (for imported data). The two numbers at the bottom of Figure 13.2 indicate that there are 3,333 rows and 22 columns of data. The first 21 columns are features, and the last column – in our case, CHURN – is the target we want to predict. In Figure 13.3, we are able to see the first few rows of the data.

Cleaning and Preparing the Data

Our first step will be data cleansing (discussed in Chapter 8). This is often a time-consuming task. Figuring out how to handle missing values is a critical part of it, since different models can be more or less sensitive to this condition, making the models more or less able or even unable to reliably predict when the available data is incomplete. This is why we first look for missing values. We can visualize them by drawing a heatmap, shown in Figure 13.4. In our example dataset, as you can see, there is some missing data in the PHONE_ NUMBER column. This is not something we will fix, because we will drop this column. However, if values are missing from more critical features, there are multiple ways to address this.

One option is to remove all rows with empty values. For instance, if the missing data for some rows is in the target column then it may be preferable to drop that entire row of data. The problem is that removing the rows risks losing critical information. A better way to deal with missing values might be to replace them with a default value derived from the dataset. In some cases, utilizing a value from the previous or following row can accomplish this. In other cases, it is done by *data imputation* – interpolating from the full dataset and using, for example, a mean. This interpolation does not always have to be the mean of numbers in this particular dataset; it could be the output of another machine learning model. In our example, we would use the sklearn framework's imputer.

The next data preparation step is to convert the categorical values into numerical values. This conversion is necessary because

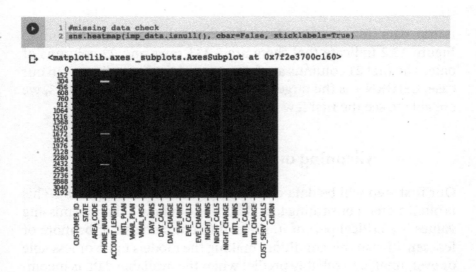

Figure 13.4 Heatmap of missing value. If there were any, they would show as a white bar for that row and column.

many machine learning models do not work with text data values. In our example, we utilize a technique called *label encoding*, which transforms yes/no and true/false values to 0 and 1 by using the sklearn LabelEncoder function (see Figure 13.5). This conversion can only be done, however, when dealing with two categories. Encoding, say, a set of state names into numerical data could introduce difficulties, since state names are categorical data with no relation of any kind to one another. Problems arise when the AI model assumes a relationship or order among assigned numbers that are in the same column – and decides, for example, that 0 is less than 1, where zero might refer to Alaska, and one might refer to Florida.

To deal with these problems, we use what is called *one-hot encoding*. One-hot encoding takes a column with label-encoded categorical data and splits it into multiple columns in which the existing numbers are replaced with 0s and 1s. In our example, we create three new columns: say, New York, California, and Michigan. For rows of customer data for which the state is New York, the column corresponding to New York is set to 1; the columns corresponding to the other states are set to 0. If the rows of customer data are for Californians, the column corresponding to

```
1  # drop features that are low impact
2  proc_data = imp_data.drop(columns=['AREA_CODE','PHONE_NUMBER'])
3  proc_data.shape
4
5  # transforming categorizal data to numerical values
6  target_features = ['INTL_PLAN', 'VMAIL_PLAN', 'CHURN']
7  for i, target_feature in enumerate(target_features):
8      print(target_feature + " : ", proc_data[target_feature].unique())
9
10 # use encoder and transform
11 encoder = LabelEncoder()
12 for i, target_feature in enumerate(target_features):
13     encoded_values = encoder.fit_transform(proc_data[target_feature].values)
14     proc_data[target_feature] = pd.Series(encoded_values, index=imp_data.index)
15     # proc_data[target_feature] = proc_data[target_feature].astype('float64')
16     print(target_feature + " : ", proc_data[target_feature].unique())
```

```
INTL_PLAN :  ['no' 'yes']
VMAIL_PLAN :  ['yes' 'no']
CHURN :  ['False.' 'True.']
INTL_PLAN :  [0 1]
VMAIL_PLAN :  [1 0]
CHURN :  [0 1]
```

Figure 13.5 Transforming categorical text data to numerical values.

```
1  # one hot encode categorical values that have more than 2 categories
2
3  proc_data = pd.get_dummies(proc_data, columns=['STATE'])
4  proc_data.shape
```

(3333, 71)

Figure 13.6   One-hot encoding of US states.

California will be 1; the columns for New York and Michigan will both be set to 0. We now use the Python get_dummies function (see Figure 13.6) to change the set of states to 51 columns (for the 50 states plus District of Columbia), with 1s in the column for each row that refers to the state represented by that column, and 0s for all the other columns. By transforming categorical values to numerical values in this fashion we can explore the data more fruitfully. You can see at the bottom of Figure 13.6 that there are now 71 columns. That is from the original 22 columns, minus the 2 columns we dropped (for area code and phone number), and the new 51 state columns we added.

Understanding the Data Using Exploratory Data Analysis

If you want to make good AI predictions, it is essential to have a good understanding of your (high-quality) training dataset. If a machine learning model fails to predict accurate future values – a not-uncommon occurrence in AI projects – it is often because of a flawed understanding of the data; the quality of results from an AI model is directly correlated with the quality of the training dataset and how well it is understood. Fixing this situation can be done iteratively, but it is easier and more effective to really understand the data, plan for the transformations needed, and then iterate through the modeling process. This level of comprehension is gained through what is known as *exploratory data analysis*. A deep understanding of the data is what often separates a good AI scientist from a mediocre one. It is neither technical nor programming-related: it involves the ability to make the right decisions about the data and choosing the most relevant model for the given situation.

One technique for doing exploratory analysis is to compute basic statistics about the data, for example, determining the mean and standard deviation of each of the features. Figure 13.7 shows the code to plot the data distribution of some of the columns. In the results, shown in Figure 13.8, we can clearly see that most of the numeric features appear to be normally distributed, although VMAIL_MSG, INT_CALLS, and CUST_SERV_CALLS are not.

Next we look at the dependencies between the features. In the correlation matrix in Figure 13.9, we can see how strongly each feature varies with other features and how the individual attributes correlate with the target attribute (last row in Figure 13.9). For example, the dark box in the cell for VMAIL_MSG (on the vertical axis) and VMAIL_PLAN (on the horizontal axis) is telling us that voicemail plan and the number of voicemail messages are highly correlated. That is understandable because you cannot receive a voicemail message if you do not have a voicemail plan. This matrix will be necessary when we are looking at feature engineering.

We can also look for outliers in our exploratory analysis, in this case, using boxplots. Each boxplot shows the median of the data column, which is shown as the line in the middle of each gray box in Figure 13.10. The gray box shows the data range from the first quartile (which is the median of the lower half of the dataset) to the third quartile (the median of the upper half of the dataset). The dots at the top and bottom are the end values that could potentially

```
1  # look at distribution of numeric data sets
2  col_names = ['ACCOUNT_LENGTH','VMAIL_MSG', 'DAY_MINS', 'DAY_CALLS', \
3              'DAY_CHARGE', 'EVE_MINS', 'EVE_CALLS', 'EVE_CHARGE', \
4              'NIGHT_MINS', 'NIGHT_CALLS', 'NIGHT_CHARGE', 'INTL_MINS', \
5              'INTL_CALLS', 'INTL_CHARGE', 'CUST_SERV_CALLS']
6
7  fig, axs = plt.subplots(5,3, figsize=(14,17))
8  for i, col_val in enumerate(col_names):
9      sns.distplot(proc_data[col_val], hist=True, ax=axs.flat[i])
10     axs.flat[i].set_xlabel(col_val, fontsize=8)
11     #axs.flat[i].set_ylabel('Count', fontsize=8)
```

Figure 13.7 Plotting frequency of datasets.

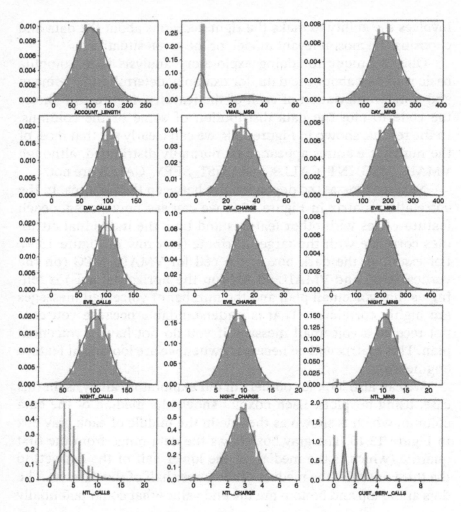

Figure 13.8 Frequency distribution of data of some of the columns.

be outliers. In our example, we will ignore the outliers in the first modeling experiment and include all the data.

An exploratory analysis also needs to look for data imbalances. In Figure 13.11, we can see that approximately 14.5% of telecom customers churned, whereas 85.5% did not. This means that if we created a "model" that simply said that, no matter the input, the

```
1  # Create data subset for visualization
2  states = ['CUSTOMER_ID', 'STATE_AK','STATE_AL','STATE_AR','STATE_AZ', \
3            'STATE_CA','STATE_CO','STATE_CT','STATE_DC','STATE_DE','STATE_FL',\
4            'STATE_GA','STATE_HI','STATE_IA','STATE_ID','STATE_IL','STATE_IN',\
5            'STATE_KS','STATE_KY','STATE_LA','STATE_MA','STATE_MD','STATE_ME',\
6            'STATE_MI','STATE_MN','STATE_MO','STATE_MS','STATE_MT','STATE_NC',\
7            'STATE_ND','STATE_NE','STATE_NH','STATE_NJ','STATE_NM','STATE_NV',\
8            'STATE_NY','STATE_OH','STATE_OK','STATE_OR','STATE_PA','STATE_RI',\
9            'STATE_SC','STATE_SD','STATE_TN','STATE_TX','STATE_UT','STATE_VA',\
10           'STATE_VT','STATE_WA','STATE_WI','STATE_WV','STATE_WY']
11 viz_data = proc_data.drop(columns=states)
12 viz_data.shape
13
14 # Compute the correlation matrix
15 corr = viz_data.corr()
16
17 # Generate a mask for the upper triangle
18 mask = np.zeros_like(corr, dtype=np.bool)
19 mask[np.triu_indices_from(mask)] = True
20
21
22 # Set up the matplotlib figure
23 f, ax = plt.subplots(figsize=(11, 9))
24
25 # Generate a custom diverging colormap
26 cmap = sns.diverging_palette(220, 10, as_cmap=True)
27
28 # Draw the heatmap with the mask and correct aspect ratio
29 sns.heatmap(corr, mask=mask, cmap=cmap, vmax=.3, center=0,
30             square=True, linewidths=.5, cbar_kws={"shrink": .5})
```

<matplotlib.axes._subplots.AxesSubplot at 0x7f2e37da4198>

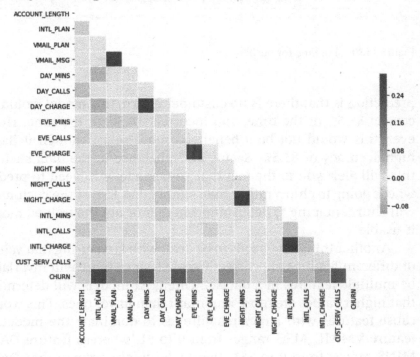

Figure 13.9   Heatmap of the correlations of some of the key columns with each other.

```
1  # exploring for outliers
2  col_names = ['ACCOUNT_LENGTH','DAY_MINS', 'DAY_CALLS', 'DAY_CHARGE', \
3               'EVE_MINS', 'EVE_CALLS', 'EVE_CHARGE', 'NIGHT_MINS', \
4               'NIGHT_CALLS', 'NIGHT_CHARGE']
5
6  fig, ax = plt.subplots(1, 10, figsize=(11,5))
7
8  for i, col_val in enumerate(col_names):
9      sns.boxplot(y=proc_data[col_val], ax=ax[i])
10     ax[i].set_ylabel('')
11     ax[i].set_xlabel(col_val, fontsize=8)
```

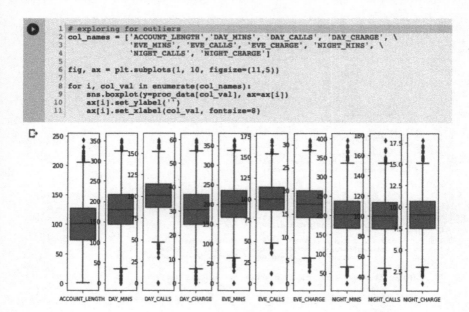

Figure 13.10 Looking for outliers.

prediction is that there is no customer churn, this model would be correct 85.5% of the time, and incorrect 14.5% of the time. However, this would not be a beneficial model even though it has a high accuracy of 85.5%. Seeing an imbalance in the data such as this will alert you to the fact that your model will need to predict who is going to churn rather than simply that 15% of the customers will churn, helping to determine whether or not the trained model is usable.

Another thing that needs to be dealt with is ensuring that values of different features are correctly scaled, since there will invariably be multiple features in a dataset. Often, AI models will determine that higher values are more important than lower ones. This would cause features with larger magnitudes to dominate the model. If feature VMAIL_MSG ranges from 0 to 51, whereas feature DAY_MINS ranges from 0 to 351, the model might assume that DAY_MINS matters more than VMAIL_MSG, which may not be the case.

```
1  # look at distribution of categorical data sets
2  num_col_names = ['INTL_PLAN','VMAIL_PLAN', 'CHURN']
3
4
5  fig, ax =plt.subplots(1, len(num_col_names), figsize=(11,6))
6  for i, col_val in enumerate(num_col_names):
7      sns.countplot(proc_data[col_val], ax=ax[i])
```

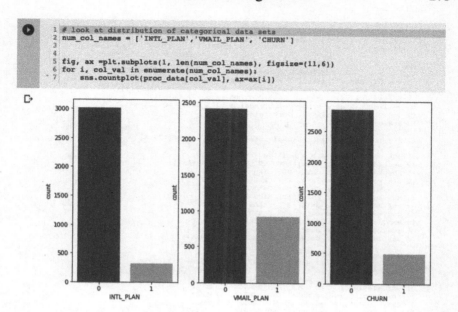

Figure 13.11   Imbalance in label or target data.

Scaling enables us to treat all features equally. As discussed in Chapter 8, there are two ways to scale features. Normalization scales the numbers to fall between zero and one, and standardization scales the number so that the values of the feature follow a normal distribution with a mean value of zero and a standard deviation of one. We use standardization (see Figure 13.12) in our example because it maintains outliers, which might contain important information that we do not want to lose.

As you can see in Figure 13.13, we are standardizing all the numeric attributes except for CUST_SERV_CALLS. We do this because satisfied customers tend not to call customer service regularly, so the number of calls to customer service might well correlate highly with churn. Since we are not scaling this feature, our model will treat it as more significant, but it will not overly dominate other feature attributes since it has a mean of approximately 1.56. If a particular feature had a higher value, such as 7, it would stand out as significant.

```
 1 # scaling the features
 2 scaler = StandardScaler()
 3 scale_cols = ['ACCOUNT_LENGTH','DAY_MINS', 'DAY_CALLS', 'DAY_CHARGE', \
 4                'EVE_MINS', 'EVE_CALLS', 'EVE_CHARGE', 'NIGHT_MINS', \
 5                'NIGHT_CALLS', 'NIGHT_CHARGE',
 6                'VMAIL_MSG', 'INTL_MINS','INTL_CALLS','INTL_CHARGE']
 7 scaled_data = scaler.fit_transform(proc_data[scale_cols])
 8 scaled_data = pd.DataFrame(scaled_data, columns=scale_cols)
 9 scaled_full_data = proc_data.drop(scale_cols, axis=1)
10 scaled_full_data = pd.concat([scaled_full_data, scaled_data], \
11                axis=1, sort=False)
12 scaled_full_data.shape
13
14 fig, (ax1, ax2) = plt.subplots(ncols=2, figsize=(12, 5))
15
16 ax1.set_title('Before Scaling')
17 sns.kdeplot(proc_data['ACCOUNT_LENGTH'], ax=ax1)
18 sns.kdeplot(proc_data['DAY_MINS'], ax=ax1)
19 sns.kdeplot(proc_data['DAY_CALLS'], ax=ax1)
20 sns.kdeplot(proc_data['DAY_CHARGE'], ax=ax1)
21 sns.kdeplot(proc_data['EVE_MINS'], ax=ax1)
22 sns.kdeplot(proc_data['EVE_CALLS'], ax=ax1)
23 sns.kdeplot(proc_data['EVE_CHARGE'], ax=ax1)
24 sns.kdeplot(proc_data['NIGHT_MINS'], ax=ax1)
25 sns.kdeplot(proc_data['NIGHT_CALLS'], ax=ax1)
26 sns.kdeplot(proc_data['NIGHT_CHARGE'], ax=ax1)
27
28 ax2.set_title('After Standard Scaler')
29 sns.kdeplot(scaled_data['ACCOUNT_LENGTH'], ax=ax2)
30 sns.kdeplot(scaled_data['DAY_MINS'], ax=ax2)
31 sns.kdeplot(scaled_data['DAY_CALLS'], ax=ax2)
32 sns.kdeplot(scaled_data['DAY_CHARGE'], ax=ax2)
33 sns.kdeplot(scaled_data['EVE_MINS'], ax=ax2)
34 sns.kdeplot(scaled_data['EVE_CALLS'], ax=ax2)
35 sns.kdeplot(scaled_data['EVE_CHARGE'], ax=ax2)
36 sns.kdeplot(scaled_data['NIGHT_MINS'], ax=ax2)
37 sns.kdeplot(scaled_data['NIGHT_CALLS'], ax=ax2)
38 sns.kdeplot(scaled_data['NIGHT_CHARGE'], ax=ax2)
39
40 plt.show()
```

Figure 13.12   Scaling the relevant data columns.

Feature Engineering

Some features are not as important as we first might think they are, while other features are significant. Adding unnecessary features can not only slow down the training but also end up overfitting the machine learning model (see Chapter 8). Moreover, keeping two different features that are highly correlated to each other can give those features too heavy an influence on the result.

From the correlation matrix in Figure 13.9, we can see which highly correlated columns we should keep as features. Based on this information, we could drop DAY_ CHARGE in favor of DAY_MINS, EVE_CHARGE in favor of EVE_MINS, NIGHT_CHARGE in favor of NIGHT_MINS, and INTL_CHARGE in favor of INTL _MINS

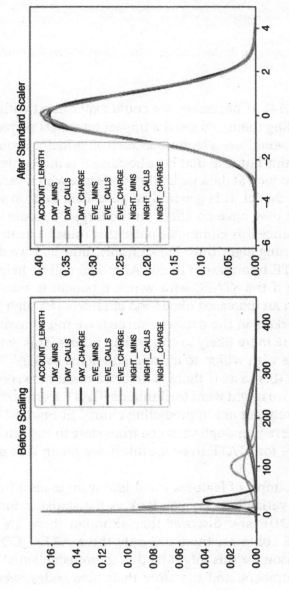

Figure 13.13 Visualizing the data distribution before scaling (left) and after scaling (right).

```
1 # remove columns with higher correlations, etc.
2 scaled_full_data['TOTAL_CHARGE'] = scaled_full_data['DAY_CHARGE'] + \
3         scaled_full_data['EVE_CHARGE'] + scaled_full_data['NIGHT_CHARGE'] + \
4         scaled_full_data['INTL_CHARGE']
5 scaled_full_data = scaled_full_data.drop(['DAY_CHARGE', 'EVE_CHARGE', \
6         'NIGHT_CHARGE', 'INTL_CHARGE'], axis = 1)
```

Figure 13.14 Dropping individual charge columns and adding the total charge column.

(see Figure 13.14). If necessary, we could experiment with including and excluding them and see the impact on model performance. Generally, we would use a technique such as principal-component analysis to do this but skip that here because it is more involved.

We can also look at data variability to decide which variables to include in the model. It is generally a good idea to drop variables if you are sure they have no effect on the target attribute – in this case, churn – since this eliminates noise that must be dealt with by the model-training algorithm. For example, how can we decide if using the STATE information in our AI model will be helpful? We can determine if the STATE with which a person is associated is correlated with an increased likelihood of churn. If a high percentage of New Yorkers in the dataset churned, we might assume that New Yorkers are more likely to churn than others. This would be a useful attribute with which to make predictions, so STATE should be included in this case. If there is no correlation between STATE and churning, we might want to eliminate STATE from our dataset, since it will likely not aid in predicting churn. In Figure 13.15, we can see that there is enough variance from state to state in churn – from 5 to 25% – for STATE to be useful, so we retain it as a feature in our dataset.

Another example of features you might want to omit from your dataset is low variability. Say you look at the statistics for STATE and AREA_CODE and discover that although there are 51 distinctive STATE columns, there are only three AREA_CODE columns. One reason for this might be that to protect clients' privacy, their phone numbers, and therefore their area codes, were omitted from the dataset. This would indicate that you should probably

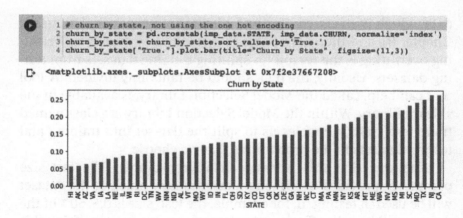

```
1  # churn by state, not using the one hot encoding
2  churn_by_state = pd.crosstab(imp_data.STATE, imp_data.CHURN, normalize='index')
3  churn_by_state = churn_by_state.sort_values(by='True.')
4  churn_by_state["True."].plot.bar(title="Churn by State", figsize=(11,3))
```

`<matplotlib.axes._subplots.AxesSubplot at 0x7f2e37667208>`

Figure 13.15 Analyzing churn rate by state.

drop the features AREA_CODE and PHONE_NUMBER (see Figure 13.5), since fictitious data will not contribute to the accuracy of your results.

Sometimes, it is useful to create new features out of one or more of those already existing in your dataset. We might, for example, decide to create a new feature by combining the four CHARGE features, since it is likely the total charge that causes someone to churn and not the charge by the time of day. We might also create a new feature that summarizes several of the original features by multiplying or dividing them. For example, we divide DAY_MINS by DAY_CALLS and come up with a new feature, DAY_AVG_CALL, the average length of a call during the day.

Creating and Selecting the Optimal Model

Models are trained on datasets, and after training, you need a different dataset on which to test your model. Recall that the goal of machine learning is to create a model with the highest predictive accuracy on data that it has not yet seen. If your training and testing dataset is the same, your model could memorize the training data, which is an extreme case of what is known as overfitting (see Chapter 8), and thus not be able to generalize to new data. Detecting overfitting is the reason for splitting training datasets from testing datasets. Usually, the data should be randomly divided. A tool that can help, called the Model Selection Library, is available in the sklearn library. Within the Model Selection Library is a class named train_test_split. It enables us to split the dataset into training and testing datasets in the proportions that we choose.

The way that this works is relatively simple. The test_size parameter, provided as a fraction, decides how much of the dataset will be used in testing. If, for example, the test_size is 0.5, 50% of the dataset will be split off as testing data. If you are not specifying this parameter, you have the option of using train_size, which operates in the same way. If you choose 0.5 as the value, 50% of the dataset will be used as the training set. If you want to determine which elements are selected for training and testing randomly, you can use the random_state parameter by choosing an integer to serve as the seed for the random number generator during the split. In our example, we are using 75% of the data for training, and 25% for testing (see Figure 13.16).

We have selected a logistic regression model as the algorithm that we want to train to be our initial AI model. Recall that when we use the term *algorithm*, we are referring to a generic algorithm, such as a linear regression algorithm. When we use the term *model*, we mean a model trained explicitly on specific data for a specific

```
1  # split the features from the target variable
2
3  sourcevars = scaled_full_data.drop(['CHURN', 'CUSTOMER_ID'], axis=1)
4  targetvar = scaled_full_data['CHURN']
5
6  # split the training and validation datasets
7
8  xTrain, xTest, yTrain, yTest = train_test_split(sourcevars, targetvar, \
9      test_size = 0.25, random_state = 0)
10
11 sourcevars.shape, targetvar.shape
```

⤷ ((3333, 65), (3333,))

Figure 13.16   Splitting data for training and testing in the ratio of 75:25.

outcome, such as our churn model. This model will be trained on historical customer churn data to predict whether a customer will leave the company based on the data inputs (features) provided. The model is based on a logistic regression algorithm, which is a general binary or multiclass classifier. Whether you choose a regression algorithm or a classification algorithm depends on the problem you are solving. If we want to predict which customers may leave, we use a classification algorithm. If we, instead, want to predict the customer lifetime value of each customer, we use a regression algorithm to get a continuous variable output.

You can see from Figure 13.17 that our model is predicting with an accuracy of 84.8%. That may seem pretty good, but recall that earlier we indicated that approximately 14.5% of the source dataset customers had churned, whereas 85.5% had not (in the full dataset of 3,333 customers). If our model predicted that no customers would ever churn, it would have an accuracy of 85.5%. Therefore, our model is actually performing worse than if it predicted no one would churn. Note that if we used only our test data (25% of the data) for prediction accuracy, we would get 86.2% (see Figure 13.18). This is slightly different from 85.5% because we are using a subset of the data to test.

Let us examine the model performance metrics more carefully. First, we look at the confusion matrix (see Figure 13.19), which is used to describe the performance of a classification model. In our case, there are two possible predicted classes – "predicted no churn" and "predicted churn" – and the data has two categories – "true no churn" and "true churn." The confusion matrix shows that 687

```
1  # try classification models
2  model = LogisticRegression(solver = 'lbfgs')
3
4  # train the algorithm on training data and predict using the testing data
5  model.fit(xTrain, yTrain)
6  predictions = model.predict(xTest)
7  print("Accuracy : ",accuracy_score(yTest, predictions, normalize = True))
```

Accuracy :   0.8477218225419664

Figure 13.17   Set up a logistic regression model for binary classification.

```
1 # this is the accuracy if you assume NO customers will churn
2 1 - yTest.mean()
```

```
0.8621103117505995
```

Figure 13.18 Percentage of customers that did not churn in the validation dataset.

true_no_churn customers were predicted accurately as pred_no_churn, but 32 actual_no_churn customers were mispredicted as pred_churn. Similarly, 20 actual_churn customers were predicted correctly as pred_churn, and 95 actual_churn customers were mispredicted as pred_no_churn. The accuracy of the model is the total correct predictions (687 + 20) as a percentage of all customers (687 + 20 + 32 + 95), which, as we saw earlier, is the not-very-useful 84.8%. To overcome this problem, we need to look at metrics beyond just accuracy.

A few other essential measures, especially in the case of imbalanced data such as ours, are called recall, precision, and F1 score. *Recall*, also called sensitivity or true positive rate, is the count of true positives (20) divided by everything actually positive (20 + 95 = 115). In our example, recall is 17% (20/115). This means we are

```
1 # print(metrics.confusion_matrix(yTest, predictions))
2 print(pd.DataFrame(confusion_matrix(yTest, predictions),
3                    columns=['pred_no_churn', 'pred_churn'],
4                    index=['actual_no_churn', 'actual_churn']))
```

	pred_no_churn	pred_churn
actual_no_churn	687	32
actual_churn	95	20

```
[173]  1 # look at performance metrics
       2 print(metrics.classification_report(yTest, predictions))
```

	precision	recall	f1-score	support
0	0.88	0.96	0.92	719
1	0.38	0.17	0.24	115
accuracy			0.85	834
macro avg	0.63	0.56	0.58	834
weighted avg	0.81	0.85	0.82	834

Figure 13.19 Looking at the confusion matrix and precision, recall, and F1 score.

accurately predicting only 17% of customers that churned. That is very low – to the point where the model could be unusable if the goal is to accurately predict which customers will churn. *Precision* is how often the prediction is correct when it has predicted yes (i.e. predicted churn). This is the count of true positives (20) divided by everything predicted to be positive (20 + 32). In our example, recall is 38% (20/52). This means that whenever we predict churn, only 38% of these predictions will be correct.

An *F1 score* is a weighted average of recall and precision: that is, recall multiplied by precision divided by recall plus precision. It represents a balance between precision and recall. The F1 score is generally used when both recall and precision are essential, and we need to find a balance between the two. In our case, the business case discussed earlier is based on higher recall, and not on the accuracy, precision, or F1 scores. Recall is more important because it will tell us which customers to try to retain. If the recall is low, we will not have identified sufficient customers to retain to make the exercise worthwhile. This metric selection is another reason why the business case should be outlined at the outset: it is essential to define what the AI model needs to do and how the outputs will be used. We now know that although we have reasonably good model accuracy, the model is inadequate to meet our business objectives.

Another way of understanding how well a model performs is to look at its receiver operating characteristic (ROC) curve (see Figure 13.20). A ROC curve is a plot of the true positive rate against the false-positive rate. An ideal model would have 100% true positives and no false positives. The area under the curve (AUC) is a measure of how good a model is. The closer the area is to 1, the better the model. In our case, the AUC is 0.81.

By applying some of the techniques explained earlier, we have determined that there are issues with the model we are using. Although it appears as if it is doing a reasonably good job of predicting how many customers will churn, it is doing a poor job of predicting which customers will churn. As you may recall, using histograms to analyze our data revealed that there was an imbalance in the churn data, which was an indication that there could be

```
1  # create ROC curve
2  plt.style.use('ggplot')
3  y_predict_probabilities = model.predict_proba(xTest)[:,1]
4
5  fpr, tpr, _ = roc_curve(yTest, y_predict_probabilities)
6  roc_auc = auc(fpr, tpr)
7
8  plt.figure()
9  plt.plot(fpr, tpr, color='darkorange', lw=2, \
10           label='ROC curve (area = %0.2f)' % roc_auc)
11 plt.plot([0, 1], [0, 1], color='navy', lw=2, linestyle='--')
12 plt.xlim([0.0, 1.0])
13 plt.ylim([0.0, 1.05])
14 plt.xlabel('False Positive Rate')
15 plt.ylabel('True Positive Rate')
16 plt.title('ROC Curve')
17 plt.legend(loc="lower right")
18 plt.show()
```

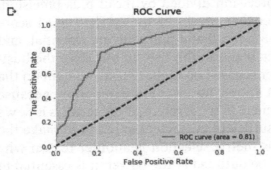

Figure 13.20 Receiver operating characteristic (ROC) curve and area under the curve (AUC).

an issue in predicting who is going to churn. To reduce this imbalance, there are a variety of techniques we can employ. One is to use *data augmentation*. In Figure 13.21, we can see the results of utilizing a method known as Synthetic Minority Over-Sampling Technique (SMOTE). By using SMOTE, we can generate additional training data for those customers who do churn. We do this by creating other samples that are similar to the underrepresented existing data – in our case, for customers who churn. We apply small, random perturbations to selected columns and use these to create new rows. When we use this data augmentation, we can see model accuracy has decreased to 73.9%, but that recall has improved significantly from 17% to 75%. Precision has dropped slightly from 38% to 31%, and the F1 score has been enhanced from 0.24 to 0.44.

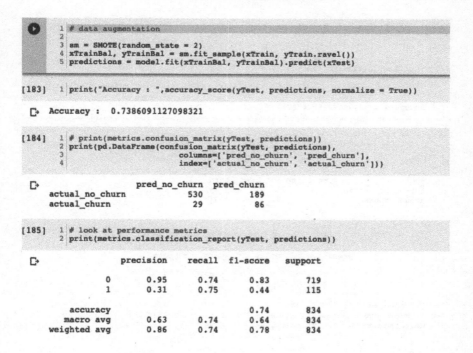

```
1 # data augmentation
2
3 sm = SMOTE(random_state = 2)
4 xTrainBal, yTrainBal = sm.fit_sample(xTrain, yTrain.ravel())
5 predictions = model.fit(xTrainBal, yTrainBal).predict(xTest)
```

```
[183]  1 print("Accuracy : ",accuracy_score(yTest, predictions, normalize = True))
```

Accuracy :   0.7386091127098321

```
[184]  1 # print(metrics.confusion_matrix(yTest, predictions))
       2 print(pd.DataFrame(confusion_matrix(yTest, predictions),
       3                    columns=['pred_no_churn', 'pred_churn'],
       4                    index=['actual_no_churn', 'actual_churn']))
```

	pred_no_churn	pred_churn
actual_no_churn	530	189
actual_churn	29	86

```
[185]  1 # look at performance metrics
       2 print(metrics.classification_report(yTest, predictions))
```

	precision	recall	f1-score	support
0	0.95	0.74	0.83	719
1	0.31	0.75	0.44	115
accuracy			0.74	834
macro avg	0.63	0.74	0.64	834
weighted avg	0.86	0.74	0.78	834

Figure 13.21   Augmenting the minority data.

It is a good idea, as discussed in Chapter 8, to try various algorithms during the modeling process. As mentioned earlier, in this example, we used a logistic regression algorithm. We could also have used a two-layer neural network or a deep neural network. We might have chosen to employ a Gaussian Process regression model, which plots all the features in terms of a Gaussian, or bell curve, enabling us to use the correlations among the features to build a predictive model, We might have decided to use an average of multiple algorithms, known as an ensemble model, to increase the accuracy of our predictions.

In our example, we will try another machine learning approach that has become popular recently: extreme gradient boosting, or XGBoost. It turns out that when we try this by changing the line of code that set our previous model to logistic regression and switching it to XGBoost Classifier, we get much better results. In Figure 13.22,

```
1 # try classification models
2 # model = LogisticRegression(solver = 'lbfgs')
3 model = xgb.XGBClassifier(objective="binary:logistic", random_state=42)
4
5 # train the algorithm on training data and predict using the testing data
6 model.fit(xTrain, yTrain)
7 predictions = model.predict(xTest)
8 print("Accuracy : ",accuracy_score(yTest, predictions, normalize = True))
```

Accuracy :   0.960431654676259

```
[189]  1 # print(metrics.confusion_matrix(yTest, predictions))
       2 print(pd.DataFrame(confusion_matrix(yTest, predictions),
       3                    columns=['pred_no_churn', 'pred_churn'],
       4                    index=['actual_no_churn', 'actual_churn']))
```

	pred_no_churn	pred_churn
actual_no_churn	710	9
actual_churn	24	91

```
[190]  1 # look at performance metrics
       2 print(metrics.classification_report(yTest, predictions))
```

	precision	recall	f1-score	support
0	0.97	0.99	0.98	719
1	0.91	0.79	0.85	115
accuracy			0.96	834
macro avg	0.94	0.89	0.91	834
weighted avg	0.96	0.96	0.96	834

Figure 13.22   Trying a different algorithm – only lines 2 and 3 in the first block have been changed to select a different model.

we can see that using XGBoost has jumped model accuracy to 96% from 74%, and improved recall slightly, from 75% to 79%. Precision has improved significantly, from 31% to 91% and, the F1 score is now at 0.85, up from 0.44. The ROC curve looks better as well, with AUC at 0.93, shown in Figure 13.23.

With this model now looking useful, we can try and understand the model a bit better by looking at what features have the most weight in the final model. In Figure 13.24, we can see the top 10 features by importance – these features have the highest impact on the predictions from the model. This model is indicating that the usage and charge are the biggest driving factors.

One other thing we might have done is to use a linear regression algorithm, which would give an output of a number between

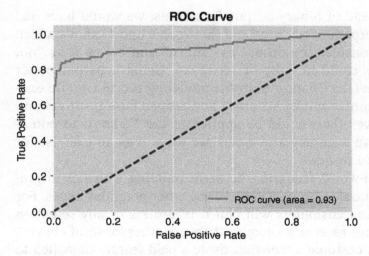

Figure 13.23   ROC curve and AUC using XGBoost.

```
1  # explore feature importance
2  fig, ax = plt.subplots(1,1,figsize=(9,6))
3  xgb.plot_importance(model, max_num_features=10, ax=ax)
4  plt.show()
```

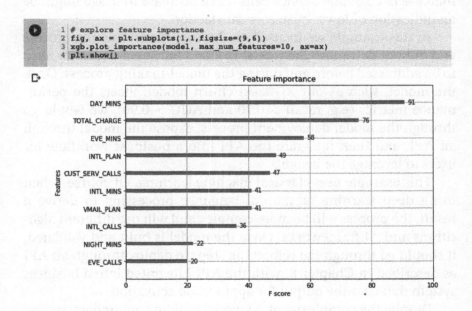

Figure 13.24   Feature importance for the top 10 features in the model.

0 and 1 instead of binary output. To do this, we would have had to set a hyperparameter threshold (between 0 and 1) above which we would consider the output as "churn" and below it as "not churn." This could initially be set at 0.5. Using hyperparameter optimization (see Chapter 8) in the modeling process would enable us to improve model accuracy and reduce false positives and false negatives. This could be applied to the XGBoost to potentially improve accuracy and recall; we leave it as an exercise for the interested reader.

In a real-world scenario, the team building a model to predict churn should also consider several other potential issues. For example, some customers will churn regardless of any retention incentives they receive. Moreover, there are other forms of churn – for example, customers who deactivate a paid feature or switch to lower-priced plans – that the team might want to model. It is a good idea to also model changes in customer behavior over time. An increase in customer service calls and a decrease in usage might be an indication of lower customer satisfaction.

In this example we focused on the core model-making process and did not delve into the model risk and fairness issues that need to be addressed before and during the model-making process. Once the model, such as our XGBoost churn model, meets the performance metrics (e.g. recall > 0.70 and AUC > 0.90), we would go through the model deployment process, expose the model through an API, and then integrate the API into a business workflow for users to leverage the insights.

This example uses classical machine learning. Of course, when using deep learning or natural language processing to derive a result, the process will be more complex and will use different algorithms and AI frameworks. Once the model is built and validated, it should go through the remaining steps to deploy through an API as described in Chapter 8, with the API integrated into a business system that uses the output for appropriate activation.

Despite the complexity of AI model building, an understanding of the process will provide managers and executives with a good sense of what their AI teams are doing and why. This will also help them make better decisions and know which metrics are relevant for what types of model outcomes.

Our next chapter will focus on the future, looking at both emerging AI technologies and some of the impacts of this now-global technology on work and society.

Note

1. Kaggle (2018). Churn in Telecoms Dataset. https://www.kaggle.com/becksddf/churn-in-telecoms-dataset/data (accessed December 16, 2019).

Part V
Looking Ahead

Chapter 14
The Future of Society, Work, and AI

The development of full artificial intelligence could spell the end of the human race.

Professor Stephen Hawking, professor of mathematics at the University of Cambridge

In 2014, the late Stephen Hawking made the above statement in a BBC News interview.[1] This somewhat terrifying quote is often thought to refer to things such as AI-enabled robots that could eventually take over the world. However, what many do not realize is what Hawking said next. "I am an optimist, and I believe that we can create AI for the good of the world," he went on. "That it can work in harmony with us. We simply need to be aware of the dangers, identify them, employ the best possible practice and management, and prepare for its consequences well in advance."

AI has the potential to improve the quality of life of everyone on the planet, raising our income levels, and helping us to live longer and healthier. It has been estimated that over 90% of all customer interactions with businesses are likely to utilize AI in some fashion in the future. Using AI, people will be able to customize the products and services they need in areas that range from banking to healthcare. AI technology will likely be deployed extensively in government agencies and legal systems around the world. AI scientists will continue to develop more sophisticated software to enable vehicles to be autonomously controlled, and AI-controlled robots and other devices will continue to grow in sophistication.

At the same time, there are a variety of challenges we will face as AI technology becomes more ubiquitous. The issue of governance, as noted in Chapter 10, is one of the critical hurdles we face as we consider the future of AI; pragmatic but effective policies and regulation are needed so that AI does not threaten human autonomy, agency, or capabilities. Regulation can also encourage a healthy, growing AI industry. Hawking himself cited some of the legislative work that was being carried out in Europe, particularly involving rules to govern AI and robotics, as being a positive development.

For the future of AI to be bright – for us to get the most from AI and further develop the technology in appropriate ways – we need to do a variety of things. We must implement existing AI technologies as well as the new ones as they emerge. We must manage the risks inherent in AI, including potential job losses; codifying bias; the malicious use of AI, such as social network manipulation through such means as deep-fakes or cyberattacks; and other unintended consequences. And we must improve and evolve AI technology through active research, both fundamental and applied.

In previous chapters, we have discussed how to apply existing AI technology to all aspects of our businesses. In this chapter, we look at the future of society and work, and how near-future developments within AI – that is, developments prior to the arrival of artificial general intelligence – will impact both the advances we expect and the challenges we will face.

AI and the Future of Society

In the years to come, AI will show up in many applications, from robotics to productivity enhancements to the emerging technology of quantum computing. In robotics, companies such as Intuitive Surgical are already developing robot-assisted technologies, tools, and services for surgical operations. Keyence and Daifuku are developing AI solutions in factory automation, and Nvidia and others are working on driverless cars.[2] There are also initiatives under way at companies such as Microsoft, Google, IBM, and others in quantum machine learning.[3]

Medical care will certainly be one of the areas in which AI will change lives. Intelligent robots will become increasingly swift and accurate and will likely be able to allow doctors to perform sophisticated robot-assisted procedures that are impossible today. AI robots will take care of the elderly; scientists are already developing a "robo-cat" who will remind geriatric patients to take their medications.

AI will aid spacecraft on their journeys to the stars and may help alleviate climate change on earth. To support developments in environmental science, Microsoft has recently committed $50 million to its newly created "AI for Earth" program.[4] The not-for-profit AI for Good Foundation[5] is another example of an enterprise that is seeking to move beyond commercial applications of AI "to help solve social, economic, and environmental problems to benefit society to the fullest extent possible."

However, although there may be debates about exactly what AI will accomplish in the future, what is not in question are the many concerns about the pervasive use of AI. Accountability is one of them. A current lawsuit over the loss of $20 million in investments due to an algorithmic "mistake"[6] is currently working its way through the courts. Unable to sue the supercomputer, the investor is trying to hold the man who sold it to him accountable for his losses. And the liability of algorithms has already been tested in Tempe, Arizona, where in March 2018, a self-driving Uber struck and killed a woman who was walking her bicycle across the street. A year later, Uber was exonerated from all criminal liability.[7] Instead, the safety driver sitting in the driver's seat is facing charges. In both cases, it was determined that the AI algorithm – and its creators – could not be held responsible.

What will happen, for example, if what is known as algorithmic governance is implemented legally, allowing, say, a police force to use surveillance on all its citizens, automatically tracking behaviors? In China, this is already occurring: China's Social Credit System (SCS) is designed to track everything from a person's Internet activity to whether she is sufficiently respectful of her parents, and an individual's ultimate rating is used to determine his or her social benefits. Although this kind of surveillance is increasing in other locations,[8] there are also places such as San Francisco that have already banned facial recognition use by the police.[9] In a world that continues to shrink, how will these opposing regulations continue to coexist?

The likelihood of AI handling all our environmental problems is also debatable. One off-the-shelf machine translation algorithm has such a massive need for computational resources that it creates a carbon footprint similar to that of the amount of fuel that five vehicles would consume during their entire lifetimes.[10] And there are other potential drawbacks to the extensive use of AI. As we rely more and more on the use of machine learning in the basic sciences, for example, we run the risk of simply predicting outcomes without truly understanding them, neglecting the development of underlying theories to explain these phenomena in the first place. In AI, this implied cost causes what Jonathan Zittrain has called the *intellectual debt* of AI.[11] It is similar to technical debt in software engineering: taking the "easy" programming route without considering better approaches that might take longer to implement but would be more useful in the future, saving on additional rework for maintenance and agility.

AI and the Future of Work

This book has already dealt with many of the positive aspects of AI in the workplace, both now and into the future. These include such things as the potential of AI to increase workplace safety by avoiding mistakes caused by human error, to eliminate repetitive and boring tasks so employees can be freed to do more creative and satisfying work, and to utilize chatbots more effectively to provide better customer service and support customer-facing teams.

The AI-driven workplace will certainly look different. Machines will interact with one another using AI algorithms and make decisions about the production chain without the need for human intervention. Even during the time of the industrial revolution, Karl Marx, in the chapter on machinery and modern industry from his book *Capital. A Critique of Political Economy*, wrote, "The instrument of labor, when it takes the form of a machine, immediately becomes a competitor of the workman himself." But those "instruments of labor" were here to stay, and it is the same with artificial intelligence. Just like the personal computer and the smartphone, AI is already becoming a part of our everyday lives. We cannot uninvent it. If we are to deal with the rise of AI successfully, it is worth remembering a few situations in which technology of one sort or another impacted the workplace, and the impacts it had as a result.

When the automobile was invented, horses were not the only ones who lost their jobs. Blacksmiths, groomers, coachmen, feed merchants, stable owners, saddlers, wheelwrights, whip makers, street cleaners, and veterinarians were among the people who lost their jobs. On the other hand, new jobs were created in automobile manufacturing plants, car dealerships, parking lots, repair establishments, and factories in which people had to create all the parts an automobile needed to function, including tires, ignitions, batteries, and carburetors.

Workers at the beginning of the twentieth century maligned factory automation, but later generations eventually profited from the two-day weekend. People viewed the advent of automated teller machines (ATMs) in the 1970s as a disaster for workers in the retail banking industry. Yet as branch costs went down, banking branch jobs increased over time, becoming less transactional in nature and more about managing customer relationships.

Similar things are happening today. There used to be 600 equity traders in the Goldman Sachs headquarters, but now there are only two.[12] That is because Goldman is using AI trading programs to automate currency and futures trading. It is also using AI to take over many of the 146 steps that go into an initial public offering. However, AI has created more jobs than those lost, and many have not so much disappeared but transformed. In fact, according to the McKinsey Global Institute, in the years before 1950, the automobile industry created 6.9 million new jobs in the United States, while 623,000 were destroyed.[13] IT advisory firm Gartner estimates that by the year 2020, AI will create 2.3 million jobs while eliminating 1.8 million.[14] And as workplaces become more efficient, humans will be free to focus on using AI devices to enhance supply chain efficiency, product development, and other tasks. This increased efficiency has the potential to lower prices, making goods available to lower-income consumers. And if the promise of AI and automation is such that the whole population has less to do, it may even be time to consider a three-day weekend.[15]

There are certainly issues that will need to be addressed along the way. Regardless of whether the net number of jobs increases or decreases, certain job classes will be deeply impacted, and although we have faced similar problems in the past, AI is causing things to happen at a faster pace. This shift will require holistic solution approaches to retraining the workforce so that the population can move toward jobs being created and away from jobs being automated. It is also easy to ignore the difficulties encountered by those whose labor enables the AI industry. Labeling training data and retraining algorithms is time-consuming and laborious. As the AI model expands to other businesses, the conditions of its workforce will likely need to be regulated to prevent the rise of AI "sweatshops."[16]

Regulating Data and Artificial Intelligence

When we think of regulation and AI, we sometimes only think in terms of algorithms. These certainly need to be regulated to protect employees and consumers. But rules about data usage are equally if not more important. Given the enormous amount of data now held privately, where we have no regulations around confidentiality or liability, there will almost certainly be problems sharing that information in the future, hampering the development of breakthrough technologies in fields such as healthcare. Perhaps even worse, collecting and using that data can lead to data abuse or other significant problems.[17]

Paul Nemitz, one of the designers of the EU's General Data Protection Regulation (GDPR), has been quoted as saying that we are moving toward "a world in which technologies like AI become all pervasive and are actually incorporating and executing the rules according to which we live in large part" and that "the absence of such framing for the internet economy has already led to a widespread culture of disregard of the law and put democracy in danger, the Facebook Cambridge Analytica scandal being only the latest wake-up call."[18]

In May 2019, 42 countries from the Organisation for Economic Co-operation and Development (OECD) agreed to a new set of policy guidelines for the development of AI systems, called "Recommendation of the Council on Artificial Intelligence."[19] It promotes five principles for the responsible development of AI:

1. Inclusive growth, sustainable development and well-being – to benefit people.
2. Human-centered values and fairness – to respect the rule of law, human rights, and democratic principles.
3. Transparency and explainability – to ensure people understand when they are interacting with AI or AI-based outcomes and can challenge them if necessary.
4. Robustness, security, and safety – to continuously assess and mitigate risks of AI throughout the AI lifecycle.
5. Accountability – to hold companies and individuals developing and deploying the technologies accountable for the proper functioning of these systems.

At the annual Summit of the Group of Seven (G7) (Canada, France, Germany, Italy, Japan, the United Kingdom, and the United States) held in France in 2018, French Prime Minister Emmanuel Macron and Canadian Prime Minister Justin Trudeau made a joint announcement about the formation of a group of international experts on AI called the International Panel on Artificial Intelligence. The panel would be modeled on the International Panel on Climate Change. More information on this panel was shared in a session in May 2019 that hosted G7 ministers of digital affairs.

One of the United States' first bills to regulate AI was introduced in April 2019 by Senators Cory Booker and Ron Wyden, with a House equivalent sponsored by Representative Yvette Clarke. Known as the Algorithmic Accountability Act, it would require the auditing of machine learning systems for bias and discrimination as well as auditing of all processes involving sensitive data. Companies will need to ensure corrective action in a reasonable time frame when such issues are discovered in these audits. Protected data would include any personally identifiable, biometric, and genetic information. The US Federal Trade Commission (FTC) would be responsible for overseeing compliance, because the FTC is also responsible for consumer protection and antitrust regulation. Another bill introduced in the United States in April 2019 would ban manipulative design practices that they allege tech giants like Facebook and Google sometimes use to get customers to give up their data.

Other countries have already drafted or passed similar legislation designed to hold technology companies legally responsible for their algorithms. Given that it is the home of both Silicon Valley and New York's Silicon Alley, the United States has a significant role to play in the international development of AI regulations, which is all the more reason for legislators and policy makers to have a deep and nuanced understanding of AI technology.

The United States is also deeply concerned about regulating disinformation and "deep-fakes," the AI technology that allows for easy creation of images and video that look real but are not – that is, visuals of an event that have never actually happened, created by manipulating images in an increasingly sophisticated way. The problem with deep-fakes is twofold. First, they can represent things that never occurred, such as embarrassing situations involving public figures that never took place. In addition, their very existence allows some people to doubt what actually happened in the past. Doubters of the 1969 moon landing, for example, might justify their beliefs based on the existence of a technology that can easily fake a trip to the moon. Some people claim that deep-fakes have been created for entertainment purposes only, and some certainly have been. But this extremely sophisticated ability to fool most of the people some of the time has chilling implications in areas such as cybercrime.

Legislation was introduced as early as 2018 to control deep-fakes, but the jury is out on whether that legislation would be successful or even enforceable. Regulations of AI technology may be driven by political agendas, so individuals must carefully craft them with a deep understanding of AI and the law. According to an article in the *Columbia Journalism Review*, the Electronic Freedom Foundation worries that a current bill working its way through Congress known as the Deepfakes Accountability Act poses some potential First Amendment problems.[20]

There are many other challenges in regulating AI. For example, there is no agreement on what AI is. Information technology is easily transportable; the data can sit in one country, the algorithm in another, and the user in a third. Will country-level regulation work? Will AI regulation need regional alliances similar to those between France and Canada cited earlier? Or will it follow recommendations made by supranational entities such as the OECD? Policies such as the June 2019 policy recommendations[21] by the American Medical Association (AMA) to ensure oversight and accountability for enhanced intelligence in healthcare may also need to be augmented by the policies of other bodies. The Global Initiative on Ethics of Autonomous and Intelligent Systems, held by the world's largest association of technical professionals, the Institute of Electrical and Electronics Engineers (IEEE), recently launched "Ethically-Aligned Design, First Edition: A Vision for Prioritizing Human Well-being with Autonomous and Intelligent Systems."[22] They call it "the most comprehensive, crowd sourced global treatise regarding the Ethics of Autonomous and Intelligent Systems available today."

Stakeholder transparency, as well as the necessity to ensure that such systems do not infringe on human rights, are only two of the issues laid out in the treatise that will need to be worked through over time. There is much work needed in this space. A recent study[23] reports, "despite an apparent agreement that AI should be 'ethical,' there is debate about both what constitutes 'ethical AI' and which ethical requirements, technical standards and best practices are needed for its realization."

The Future of AI: Improving AI Technology

Economic growth is driven by technological innovation, particularly the creation of what has been dubbed general technology: innovations such as the ability to generate steam and electrical power and the internal combustion engine. AI is the latest to be considered a general technology. To advance its use, we need to continue to improve and evolve the technology itself through fundamental and applied research.

As the capabilities of AI evolve, it will find more applications within businesses and in our daily lives. Among the types of AI that may be widely used in the future are reinforcement learning, generative adversarial learning, federated learning, natural language processing, capsule networks, and quantum machine learning.

Reinforcement Learning

In addition to supervised, unsupervised, or semisupervised machine learning (see Chapter 2), there is another approach: *reinforcement learning* (RL). Reinforcement learning resembles what Pavlov did when training rats. As Pavlov rewarded his animals with food pellets when they successfully navigated a maze, optimal action or behavior in machine learning systems is reinforced with rewards – namely, numerical values with which they are credited for taking a particular action at a specific time. The algorithm is tasked with trying different sequences of actions to find the optimal one that will maximize its reward. This reward is tied to the success of some objective that the algorithm is trying to accomplish – for example, enabling a robot to walk successfully or a car to drive safely.

What makes this particularly interesting is that reinforcement learning does not require preexisting data to create models. Using a handful of instructions, it enables a computer to analyze a situation and then generate data by trial and error based on that situation. If the problem is complex, a reinforcement-learning algorithm can adapt over time, if necessary, to increase its rewards. However, it is difficult to use reinforcement learning successfully unless problems have a clear and quantifiable rewards structure, and the environments that RL will operate in are easily described.

Deep reinforcement learning uses deep neural networks together with the framework of reinforcements to achieve almost human-level performance in certain activities.[24] Google has reported that using this technology, computers have learned to play games on the Atari 2600 console using a system of rewards for good gameplay. The algorithm achieved human-like performance in almost half of all the games to which it was applied. Reinforcement learning became popular due to the successes of Google DeepMind's AlphaGo and AlphaZero systems. The reason RL has mostly been successful in areas involving games and simulations is that in these environments, it can learn easily by trial and error. If, instead of a simulation, it was interacting with actual customers, patients, or physical machines, this would be a very expensive way to learn.

Reinforcement learning is an active part of research, with commercial uses only recently emerging. Its utility will continue to increase as companies begin to utilize it in specific applications within situations in which historical labeled data is not readily available. It has powerful real-world implications as well. Google used a similar model to reduce power consumption by 40% at its data centers.[25] In other companies, two areas in which RL has been applied are in process configuration and sequential decision-making. For example, by analyzing and optimizing sequences, RL can determine optimal traffic configurations, dictating precisely when and where to change traffic signals. This model is also used to optimize resources at data centers and control network traffic. Ultimately, the hope is that RL will be able to solve a wide variety of business problems. To get there, however, business applications would require scenarios in which the trial and error is low cost, ideally using a realistic simulation environment to train and test the RL agent.

Generative Adversarial Learning

Recently, a new form of machine learning has been developed that is able to generate new images, speech, or text that are indistinguishable from actual images in training data. This form of machine learning is known as a generative adversarial learning. The models that are used are called *generative adversarial networks* (GANs).

GANs, first introduced by Ian Goodfellow in 2014, are particularly useful because they offer a new way to do unsupervised learning. They consist of two neural networks – a generator that takes input and produces new samples, and a discriminator that learns what the real input looks like and is tasked with distinguishing the real from the fake. When we use discriminative models for classification or regression learning, the technique employed involves learning from labeled samples and looking at unlabeled samples to decide what the labels for these are. Alternatively, a discriminative model could decide whether these unlabeled samples are part of the labeled class or not. Generative models, on the other hand, use training data and learn to generate data that appears similar to this training data. It does this by determining which combinations of features makes a sample appear to be similar. The challenge, which is often represented as a game, is for the generator to fool the discriminator into thinking the data it generates is real, and for the discriminator to successfully distinguish one from the other (hence the name "adversarial"). Over time, the generator gets better at performing its task and can fool the discriminator. A GAN model, for example, can take training data composed of images of faces and produce synthesized images that resemble real faces, although the people they are supposed to represent do not actually exist.[26] There are a variety of academic applications using GANs, and the technology is progressing rapidly, but thus far, broader business applications have yet to emerge.

Federated Learning

Typical machine learning applications require that data be collected either in one computer or at a data center – that is, in a relatively small and centralized environment. Edge computing allows us to process and analyze data near its source. When data is used locally at the edge by devices – such as smart thermostats – it requires less data flow, reducing network traffic and thus response time.

Federated learning is a term that refers to the ability of Internet of Things (IoT) devices, such as mobile phones or drones, to share their insights without sharing their data through the cloud. This enables edge devices to train models instead of needing pretrained models. In a sense, federated learning brings machine learning to edge computing in a way that has not existed before. The model – say, your thermostat – downloads the latest software and then learns from data it collects. An encrypted model it has learned is the only thing that it uploads to the cloud. The original data never leaves your thermostat. Then these learned models from each thermostat are aggregated (for example, by averaging) into one model, and the new model is redistributed to the edge. In this iterative way, the model improves over time.

Federated learning will have multiple applications in the future, playing an important role in a variety of areas. Mobile devices will require less time to generate new insights and act on them. Security will improve, since insights can be shared without sharing the data from which the insights are generated. Data privacy will be served, since users will no longer have to send sensitive information to various companies over networks owned by service providers.

Efficiency will increase as edge devices take on the tasks for which they are best suited, and it will enable more private and cost-effective machine learning applications.

Federated learning is likely to become more important as the data privacy concerns of individual users become more prevalent and mobile and edge devices become more powerful.

Natural Language Processing

Natural language processing (NLP) enables computers to extract and analyze information from natural language text and then answer questions, retrieve information, generate text, and translate one language to another. Since a major goal of AI is to enable computers and smart devices to understand and apply spoken and written languages as well as solve problems, NLP has become an important area of research. There is so much unstructured text data in the world that parsing through it quickly and easily has become imperative. In addition, small smart devices such as cell phones encourage the use of natural language as the easiest and most intuitive form of input and output.

A major hurdle facing NLP is the variety of meanings that language can have. The current approach to NLP involves representing words and text as vectors, each one of which is a set of real numbers. This allows for capturing a word's relationship to other words, what words it appears with or next to, and with what frequency. These vectors, called *word embeddings*, support language translation and information search and retrieval. Using word embeddings does have some drawbacks, however. One problem is that it has little sensitivity to context; wherever a word appears, its representation is the same. Computer scientists have recently used a neural sequence encoder to add contextual information, which has greatly improved traditional word embedding.

In just the past few years, major strides have enabled language models that not only support data mining, but also simulate very rudimentary types of reasoning, allowing these programs to understand the material and present them in a meaningful way to humans. These advances recently helped an AI system called AristoBERT pass an eighth-grade science test.[27] Even these recent advances, however, have not enabled NLP technologies to be at the same level of performance as are current computer vision or image recognition technologies. NLP remains a very active field of research, with investigations proceeding in the areas of new neural network structures, transfer learning, how to utilize knowledge and common sense in natural language understanding, and data augmentation methods such as introducing domain knowledge or using dictionaries and synonyms to improve performance.

Capsule Networks

Deep neural networks need gigantic quantities of data to train on. In many cases, this data may not be available or may be cost prohibitive to acquire. This challenge led Google's Geoffrey Hinton and his students to the idea of *capsule networks*.[28] Capsule networks utilize small groups of neurons, known as capsules, to better model hierarchical relationships. One of the problems capsule networks address is sometimes referred to as the "Picasso problem" in image recognition. If an image of a human face is cut up and collaged back (like a Picasso painting of a face), a deep neural network will likely still classify it as a "human" face, because even with large amounts of training data, it does not fully encapsulate the hierarchical relationships of nose, eyes, mouth, eyebrows, face, hair, and head.

A capsule network on the other hand, would be able to recognize this as not a face, with much less training data, because it learns the hierarchical relationships of the parts of the head and the head itself, whereas it would still be able to recognize images of the head from different angles or points of view. Currently, these capsule networks perform a bit more slowly than traditional neural nets, but Hinton suggests that they may eventually provide a way to more efficiently solve problems using less data to train on.

Quantum Machine Learning

Quantum computing is a bleeding-edge technology that, if successful, could radically transform not only the speed at which certain calculations are done, but also the computational paradigm itself. It is based on quantum mechanics: the latest and most accepted theory of how the physical world works, which enables much of modern technology, from computer chips to DVD players to nuclear power plants. Developed by Erwin Schrödinger and Werner Heisenberg, quantum mechanics was subsequently popularized by Schrödinger in his description of a cat in a box that could simultaneously be both dead (or zero) and alive (one).

Computing today is based on bits; those bits have definite values of one or zero. In quantum computing, instead of a bit being either definitively one or zero, it is a combination of both – like Schrödinger's cat. Quantum computing calls these quantum bits, or qubits, and the interesting thing about them is that they can exist in multiple states at the same time. To operate in this mysterious quantum realm, molecules are chilled to near absolute zero Kelvin, which is colder than it is in deep space. When qubits are then entangled, even if they are not near one another, they still affect one another's behavior.

Quantum computers use quantum effects, such as quantum coherence, to process information. Quantum computing can theoretically do far better than current computers in solving several problems, including searching an unsorted database and inverting a sparse matrix. Currently, it is only taking place in the lab at scales involving very few qubits, and only certain problems are amenable to it. Google recently published a paper indicating that their quantum processor "takes about 200 seconds" to complete a task and that their benchmarks indicate that "the equivalent task for a state-of-the-art classical supercomputer would take approximately 10,000 years."[29] If experiments continue to be successful, its impact on AI, and computation in general, will be profound. The idea of using quantum computing for machine learning is an emerging area of interest. Already, a team from IBM performed a simple machine learning test, first without entangling the qubits, and then with them entangled. In the first test, the error rate was 5%; in the second, the error rate was 2.5%.[30] This may not be a significant difference, but it is an indicator that quantum computing may transform AI in the future.

And This Is Just the Beginning

Given the intensity of focus on artificial intelligence, more progress will continue to be made in fundamental research, applied research, and improved toolsets, and AI will continue to be employed for a host of additional use cases. There is certainly enough interest in the field – I recently visited an advanced PhD level class on deep learning in which 120 students had enrolled – an enormous number for any graduate department.

In the future, AI will be made less data dependent, more transparent and interpretable,[31] and less biased. Areas such as capsule networks and knowledge modeling will continue to grow, and more sophisticated AI solutions will be developed in conjunction with other technologies, such as robotics, quantum machine learning, augmented reality, and IoT. It will be more frequently used for high-velocity decision-making and move toward autonomous operations. This is not only where the output of the AI models will be shown on a report or integrated into an application for human consumption and decision-making, but where the AI will sense what is happening, make the decisions, and take actions based on it. We might find ourselves enjoying those three-day weekends by as early as 2030, with better health and higher standards of living for everyone because of AI.[32]

Many areas involving AI technology will be fraught with challenges that we will have to overcome as both business and technology leaders and as citizens. It will require many adjustments, but it is poised to drive the largest business transformation in our history. We are extremely lucky to be working at a time when our use of AI can have this kind of outsized positive impact on businesses and on our lives. It is time to jump in and start adapting your organization to take advantage of it.

Notes

1. BBC News (December 2, 2014). Stephen Hawking Warns Artificial Intelligence Could End Mankind. https://www.bbc.com/news/technology-30290540 (accessed September 26, 2019).

2. *US News* (May 7, 2018). Robotics, Automation and AI Are the New FANG. https://money.usnews.com/investing/investing-101/articles/2018-05-07/robotics-automation-and-ai-are-the-new-fang (accessed September 26, 2019).

3. The Next Web (2018). New Physics AI Could Be the Key to a Quantum Computing Revolution. https://thenextweb.com/artificial-intelligence/2018/09/19/new-physics-ai-could-be-the-key-to-a-quantum-computing-revolution/ (accessed September 26, 2019).

4. *MIT Technology Review* (December 11, 2017). Microsoft Announces $50 Million for Its "AI for Earth" Project. https://www.technologyreview.com/f/609745/microsoft-announces-50-million-for-its-ai-for-earth-project/ (accessed September 26, 2019).

5. AI for Good Foundation. How Can AI and Machine Learning Be Applied to Solve Some of Society's Biggest Challenges? https://ai4good.org (accessed September 26, 2019).

6. Bloomberg (May 5, 2019). Who to Sue When a Robot Loses Your Fortune. https://www.bloomberg.com/news/articles/2019-05-06/who-to-sue-when-a-robot-loses-your-fortune (accessed September 26, 2019).

7. *New Tork Times* (March 5, 2019). Prosecutors Don't Plan to Charge Uber in Self-Driving Car's Fatal Accident. https://www.nytimes.com/2019/03/05/technology/uber-self-driving-car-arizona.html (accessed September 26, 2019).

8. *New York Times* (April 24, 2019). Made in China, Exported to the World: The Surveillance State. https://www.nytimes.com/2019/04/24/technology/ecuador-surveillance-cameras-police-government.html (accessed September 26, 2019).

9. *New York Times* (May 14, 2019). San Francisco Bans Facial Recognition Technology. https://www.nytimes.com/2019/05/14/us/facial-recognition-ban-san-francisco.html (accessed September 26, 2019).

10. *MIT Technology Review* (June 6, 2019). Training a Single AI Model Can Emit as Much Carbon as Five Cars in Their Lifetimes. https://www.technologyreview.com/s/613630/training-a-single-ai-model-can-emit-as-much-carbon-as-five-cars-in-their-lifetimes/ (accessed September 26, 2019).

11. Boing Boing (July 28, 2019). Intellectual Debt: It's Bad Enough When AI Gets Its Predictions Wrong, but It's Potentially WORSE When AI Gets It Right. https://boingboing.net/2019/07/28/orphans-of-the-sky.html (accessed September 26, 2019).

12. *MIT Technology Review* (February 7, 2019). As Goldman Embraces Automation, Even the Masters of the Universe Are Threatened. https://www.technologyreview.com/s/603431/as-goldman-embraces-automation-even-the-masters-of-the-universe-are-threatened/ (accessed September 26, 2019).

13. McKinsey Global Institute (December 1, 2017). Jobs Lost, Jobs Gained: Workforce Transitions in a Time of Automation. https://www.mckinsey.com/~/media/mckinsey/featured%20insights/future%20of%20organizations/what%20the%20future%20of%20work%20will%20mean%20for%20jobs%20skills%20and%20wages/mgi-jobs-lost-jobs-gained-report-december-6-2017.ashx (accessed September 26, 2019).

14. Gartner. Future-Proof Your Talent Strategy. https://www.gartner.com/en/human-resources/research-tools/talentneuron/future-proof-your-talent-strategy (accessed September 26, 2019).

15. *The Economist* (September 21, 2018). Why the Weekend Isn't Longer. https://www.economist.com/the-economist-explains/2018/09/21/why-the-weekend-isnt-longer (accessed September 26, 2019).

16. *The Guardian* (June 25, 2019). A White-collar Sweatshop: Google Assistant Contractors Allege Wage Theft. https://www.theguardian.com/technology/2019/may/28/a-white-collar-sweatshop-google-assistant-contractors-allege-wage-theft (accessed September 26, 2019).

17. The documentary *The Great Hack* (2019) examines the Cambridge Analytica scandal.

18. Philosophical Transactions of the Royal Society (October 15, 2018). Constitutional Democracy and Technology in the Age of Artificial Intelligence. https://royalsocietypublishing.org/doi/full/10.1098/rsta.2018.0089 (accessed September 26, 2019).

19. Organisation for Economic Co-operation and Development (May 21, 2019). Recommendation of the Council on Artificial Intelligence. https://legalinstruments.oecd.org/en/instruments/OECD-LEAL-0449 (accessed September 26, 2019).

20. *Columbia Journalism Review* (July 1, 2019). Legislation Aimed at Stopping Deepfakes Is a Bad Idea. https://www.cjr.org/analysis/legislation-deepfakes.php (accessed September 26, 2019).

21. American Medical Association (June 14, 2018). AMA Passes First Policy Recommendations on Augmented Intelligence. https://www.ama-assn.org/press-center/press-releases/ama-passes-first-policy-recommendations-augmented-intelligence (accessed September 26, 2019).

22. Institute of Electrical and Electronics Engineers (December 2017). Ethically Aligned Design. https://ethicsinaction.ieee.org (accessed September 26, 2019).

23. Nature Machine Intelligence (September 2, 2019). The Global Landscape of AI Ethics Guidelines. https://www.nature.com/articles/s42256-019-0088-2 (accessed September 26, 2019).

24. Google DeepMind (June 17, 2016). Deep Reinforcement Learning. https://deepmind.com/blog/deep-reinforcement-learning/ (accessed September 26, 2019).

25. Google DeepMind (July 20, 2016). DeepMind AI Reduces Google Data Centre Cooling Bill by 40%. https://deepmind.com/blog/article/deepmind-ai-reduces-google-data-centre-cooling-bill-40 (accessed September 26, 2019).

26. Images of people are generated by GANs at https://thispersondoesnotexist.com – these people do not actually exist (accessed September 26, 2019).

27. *New York Times* (September 4, 2019). A Breakthrough for A.I. Technology: Passing an 8th-Grade Science Test. https://www.nytimes.com/2019/09/04/technology/artificial-intelligence-aristo-passed-test.html (accessed September 26, 2019).

28. *MIT Technology Review* (November 1, 2017). Google Researchers Have a New Alternative to Traditional Neural Networks. https://www.technologyreview.com/the-download/609297/google-researchers-have-a-new-alternative-to-traditional-neural-networks/ (accessed September 26, 2019).

29. *Nature* (October 23, 2019). Quantum Supremacy Using a Programmable Superconducting Processor. https://www.nature.com/articles/s41586-019-1666-5 (accessed December 16, 2019).

30. *MIT Technology Review* (March 26, 2018). IBM's Dario Gil Says Quantum Computing Promises to Accelerate AI. https://www.technologyreview.com/s/610624/ibms-dario-gil-says-quantum-computing-promises-to-accelerate-ai/ (accessed September 26, 2019).

31. *Wired* (October 8, 2019). An AI Pioneer Wants His Algorithms to Understand the "'Why." https://www.wired.com/story/ai-pioneer-algorithms-understand-why/ (accessed December 16, 2019).

32. Inc.com (November 4, 2019). It's Official: We Should All Stop Working on Fridays, According to Microsoft. https://www.inc.com/john-brandon/its-official-we-should-all-stop-working-on-fridays-according-to-microsoft.html (accessed December 16, 2019).

Further Reading

General

Barrat, J. (2013). *Our Final Invention: Artificial Intelligence and the End of the Human Era.* New York: Thomas Dunne.

Broussard, M. (2018). *Artificial Unintelligence: How Computers Misunderstand the World.* Cambridge: MIT Press.

Domingo, P. (2015). *The Master Algorithm: How the Quest for the Ultimate Learning Machine Will Remake Our World.* New York: Basic Books.

Harari, Y. N. (2015). *Sapiens: A Brief History of Humankind.* New York: HarperCollins.

Harari, Y. N. (2017). *Homo Deus: A Brief History of Tomorrow.* New York: HarperCollins.

Kurzweil, R. (2005). *The Singularity Is Near: When Humans Transcend Biology.* New York: Penguin Group.

Lee, K. (2018). *AI Superpowers: China, Silicon Valley, and the New World Order.* New York: Houghton Mifflin Harcourt.

Maeda, J. (2019). *How to Speak Machine: Computational Thinking for the Rest of Us.* London: Portfolio.

Mitchell, M. (2019). *Artificial Intelligence: A Guide for Thinking Humans.* New York: Farrar, Straus and Giroux.

313

Page, S. E. (2018). *The Model Thinker: What You Need to Know to Make Data Work for You.* New York: Basic Books.

Rosling, H., O. Rosling, and A. Rönlund. (2018). *Factfulness, Ten Reasons We're Wrong About the World – and Why Things Are Better Than You Think.* New York: Flatiron Books.

Sejnowski, T. (2018). *The Deep Learning Revolution.* Cambridge, MA: MIT Press.

Tegmark, M. (2014). *Our Mathematical Universe: My Quest for the Ultimate Nature of Reality.* New York: Knopf.

Wachter-Boettcher, S. (2017). *Technically Wrong: Sexist Apps, Biased Algorithms, and Other Threats of Toxic Tech.* New York: W.W. Norton & Company.

Society

Bostrom, N. (2014). *Superintelligence: Paths, Dangers, Strategies.* Oxford, UK: Oxford University Press.

Cheney-Lippold, J. (2017). *We Are Data: Algorithms and the Making of Our Digital Selves.* New York: New York University Press.

Christian, B., and T. Griffiths. (2016). *Algorithms to Live By: The Computer Science of Human Decisions.* New York: Picador.

Goodman, M. (2016). *Future Crimes: Inside the Digital Underground and the Battle for Our Connected World.* New York: Knopf.

Holt, T. J., A. M. Bossler, and K. C. Seigried-Spellar. (2015). *Cybercrime and Digital Forensics: An Introduction.* New York: Routledge.

Husain, A. (2017). *The Sentient Machine: The Coming Age of Artificial Intelligence.* New York: Simon & Schuster.

Marcus, G., and E. Davis. (2019). *Rebooting AI: Building Artificial Intelligence We Can Trust.* New York: Pantheon Books.

Rotenberg, M. (2019). *The AI Policy Sourcebook 2019.* Washington, DC: EPIC.

Scharf, R. (2019). *Alexa Is Stealing Your Job: The Impact of Artificial Intelligence on Your Future.* New York: Morgan James.

Tegmark, M. (2017). *Life 3.0: Being Human in the Age of Artificial Intelligence.* New York: Knopf.

Turner, R. (2019). *Robot Rules: Regulating Artificial Intelligence.* Cham, Switzerland: Palgrave Macmillan.

Walsh, M. (2019). *The Algorithmic Leader: How to Be Smarter When Machines Are Smart Than You.* Canada: Page Two Books.

Zuboff, S. (2019). *The Age of Surveillance Capitalism: The Fight for a Human Future at the New Frontier of Power.* New York: Hachette Book Group.

Work

Agarwal, A., J. Gans, and A. Goldfarb. (2018). *Prediction Machines: The Simple Economics of Artificial Intelligence*. Boston: Harvard Business Review Press.

Ashley, K. D. (2017). *Artificial Intelligence and Legal Analytics: New Tools for Law Practice in the Digital Age*. Cambridge, UK: Cambridge University Press.

Burmeister, C. (2019). AI for Sales: How Artificial Intelligence Is Changing Sales. Chad Burmeister.

Finlay, S. (2017). *Artificial Intelligence and Machine Learning for Business: A No-Nonsense Guide to Data Driven Technologies*. Lancashire: Relativistic Books.

Guida, T. (2019). *Big Data and Machine Learning in Quantitative Investments*. West Sussex, UK: Wiley.

Marr, B. (2017). *Data Strategy: How to Profit from a World of Big Data, Analytics and the Internet of Things*. London: Kogan Page Limited.

Mather, B. (2018). *Artificial Intelligence Business Applications: Artificial Intelligence and Sales Applications*. Seattle: Amazon Digital Services LLC.

Panesar, A. (2019). *Machine Learning and AI for Healthcare: Big Data for Improved Health Outcomes*. New York: Apress.

Provost, F. and T. Fawcett. (2013). *Data Science for Business: What You Need to Know About Data Mining and Data-Analytic Thinking*. Sebastopol, CA: O'Reilly Media.

Richardson, V. J., R. A. Teeter, and K. L.Terrell. (2018). *Data Analytics for Accounting*. New York: McGraw-Hill Education.

Sterne, J. (2017). *Artificial Intelligence for Marketing: Practical Applications*. Hoboken, NJ: Wiley.

Topol, E. (2019). *Deep Medicine: How Artificial Intelligence Can Make Healthcare Human Again*. New York: Basic Books.

Ward, C. J., and J. J. Ward. (2019). *Data Leverage: Unlocking the Surprising Growth Potential of Data Partnerships*. Miami: Ward PLLC.

Yao, M., M. Jai, and A. Zhou. (2018). *Applied Artificial Intelligence: A Handbook for Business Leaders*. New York: Topbots.

Acknowledgments

O ver the years, many people have inspired me both at work and in my home life, and either directly or indirectly helped to shape this book.

I am grateful to Randall Orbon and Adriana Miller, whose enthusiasm made this book possible.

I am deeply appreciative of all the editors who helped me to refine my ideas, sharpen my language, and create a more compelling narrative: Kevin Harreld, Elisha Benjamin, Pilar Patton, Bill Kozel, and Louise Gikow.

Without the support of my peers, teams, clients, and partners at Publicis Sapient, as well as the people who took the time to educate me and share their perspectives, I would not have arrived at the point where I could write this book. For this I want to thank Sray Agarwal, Nitin Agrawal, Professor Genevera Allen, Hilding Anderson, Rohit Arora, Irakli Beridze, Kanishka Bhattacharya, Bill Braun, Hugh Connett, Art Crosby, Rodney Coutinho, Milind Godbole, Steve Guggenheimer, Shahed Haq, Professor Catherine Havasi,

Jeremy Howard, Professor Chris Jermaine, Cassie Kozyrkov, Dan Lambright, Tim Lawless, Burton McFarland, Hugo Manessi, David Murphy, Andrew Ng, Satyendra Pal, Professor Ankit Patel, David Poole, Amit Singh, Ahsan Sohail, Josh Sutton, Cobus Theunissen, Kevin Troyanos, Ashish Tyagi, Bob Van Beber, and Ray Velez.

I continue to be inspired by the people who ignited my initial interest in models and algorithms: Masud Haq, Nandini Ramachandran, Saly Kutty Joseph, Tayyba Kanwal, Farhad Faisal, Joseph Eberly, Joseph Niesendorfer, Stephen Hawking, John Archibald Wheeler, Roger Penrose, Howard Carmichael, and Jim Isenberg.

I am thankful to my mother, Raushan Hasina Haq, for showing me (among many other things) that it is possible to actually write a book!

Finally, I am deeply grateful to my family for putting up with my absences with good humor over the past two years while I was writing this book: Athena Haq, Darius Haq, and Tayyba Kanwal.

About the Author

Rashed Haq is an American technologist, scientist and artist. He was recently appointed as the Vice President of Robotics at Cruise, one of the leading autonomous vehicle companies. He was previously the Global Head of AI & Data and Group Vice President at Publicis Sapient. An accomplished analytics and technology visionary, he has spent over 20 years helping companies transform and create sustained competitive advantage through innovative applications of artificial intelligence, dynamic optimization, advanced analytics, and big data engineering. With an eye toward the future and what's possible at the intersection of technology, business, data, and algorithms, Rashed has spearheaded advanced analytics work to help companies create new products and services, generate revenue, cut costs, and reduce risk.

Rashed holds advanced degrees in theoretical physics and mathematics. Prior to Sapient, he conducted research in physics at the Los Alamos National Lab and the Institute for Theoretical Science. He also worked with Silicon Valley companies designing complex algorithms and implementing the Internet's first web translation application.

An accomplished author and sought-after speaker, Rashed frequently writes about the practical uses of AI in business and speaks about the application of safe AI and analytics at global conferences, such as AI Summit, AI in Finance, AI Pioneers Forum, LEAP Energy, Energy Risk, EMART, and the Asia Chamber of Commerce. He serves on the AI Advisory Board of the Computing Technology Industry Association.

Rashed lives in Houston with his wife and two children. When he's not focused on the future of business, technology, and science, Rashed enjoys creating art, having shown his work at many gallery exhibitions across North America.

Index

Note: Page references in *italics* refer to figures and tables.